# Precision Castin

**Pergamon Titles of Related Interest**

ASHBY & JONES
Engineering Materials 1 & 2

HEARN
Mechanics of Materials, 2-Volume Set, 2nd Edition

IIW
Joining/Welding 2000

KELLY
Concise Encyclopedia of Composite Materials

OSGOOD
Fatigue Design, 2nd Edition

O'SHIMA & VAN RIJN
Production Control in Process Industry

TAYA & ARSENAULT
Metal Matrix Composites

ZHANG
Theory and Technique of Precision Cutting

**Pergamon Related Journals**

*(free specimen copy gladly sent on request)*

Acta Metallurgica et Materialia

Engineering Fracture Mechanics

Fatigue and Fracture of Engineering Materials & Structures

International Journal of Engineering Science

International Journal of Machine Tools and Manufacture

International Journal of Mechanical Sciences

Mechanism and Machine Theory

# Precision Casting Processes

A. J. CLEGG
*Loughborough University of Technology, Leicestershire, UK*

**PERGAMON PRESS**
OXFORD · NEW YORK · BEIJING · FRANKFURT
SEOUL · SYDNEY · TOKYO

| | |
|---|---|
| UK | Pergamon Press plc, Headington Hill Hall, Oxford OX3 0BW, England |
| USA | Pergamon Press, Inc, Maxwell House, Fairview Park, Elmsford, New York 10523, USA |
| PEOPLE'S REPUBLIC OF CHINA | Maxwell Pergamon China, Beijing Exhibition Centre, Xizhimenwai Dajie, Beijing 100044, People's Republic of China |
| GERMANY | Pergamon Press GmbH, Hammerweg 6, D-6242 Kronberg, Germany |
| KOREA | Pergamon Press Korea, KPO Box 315, Seoul 110-603, Korea |
| AUSTRALIA | Maxwell Macmillan Pergamon Publishing Australia Pty Ltd, Lakes Business Park, 2 Lord Street, Botany, NSW 2019, Australia |
| JAPAN | Pergamon Press, 8th Floor, Matsuoka Central Building, 1-7-1 Nishi-Shinjuku, Shinjuku-ku, Tokyo 160, Japan |



First edition 1991

**Library of Congress Cataloging-in-Publication Data**

Clegg, A. J.
Precision casting processes/A. J. Clegg.—1st ed.
p. cm.
Includes index.
1. Precision casting. I. Title.
TS233.C54 1991 671.2'55—dc20 91-19407

**British Library Cataloguing in Publication Data**

A catalogue record for this title is available from the British Library.

ISBN 0-08-037878-1 Flexicover
ISBN 0-08-037879-X Hardcover

*Printed in Great Britain by BPCC Wheatons Ltd, Exeter*

# Contents

# Preface

The casting process is without equal as the manufacturing process capable of producing fully shaped components of any size in practically any metal or alloy. Casting users are placing increasing demands on the foundry industry to produce castings of improved quality. This is reflected in the demands for improved dimensional accuracy and consistency, improved surface finish and improved metallurgical integrity. These qualities are the recognised attributes of precision castings but what are precision castings, how are they defined and how can they be made? The objective of this book is to describe the principal casting production processes available to the foundry industry and to review the factors influencing their ability to produce precision castings. The book will be of interest to practising foundrymen, to designers and users of castings, and to students of engineering and materials.

The main emphasis of the book is therefore within Chapters 2 to 8 which provide descriptions of the processes, their capabilities and the materials used. However, Chapter 1 provides an introduction to the casting process and the foundry industry and defines the term precision in the context of casting production. Whilst the emphasis of the book is placed upon the casting processes, the book would be incomplete if it did not recognise the importance of casting design, metal melting and treatment, and quality control procedures in the production of precision castings. These aspects are outlined in Chapter 9, which considers the relationship between the casting process and design constraints, in Chapter 10, which identifies the importance of metal melting and treatment in sound casting production, and in Chapter 11, which identifies the procedures used by the foundry industry to assure its customers of their castings' quality. References are provided at the end of each chapter, both for further reading and as an acknowledgement of the source of the information reviewed and used in the chapter.

I am grateful to a number of people and organisations for their help:

The Directors of FMJ International Publications Ltd, and especially Mr C J McCombe, for permission to include extracts from papers and articles published under my authorship in the company's journals and for the provision of many of the photographs reproduced in the text.

The Directors of the following organisations for their permission to use illustrative material:

| | |
|---|---|
| BCIRA | Figs. 2.4, 2.5, 2.6, 2.7, 2.17 |
| Borden (UK) Ltd | Figs. 3.1, 3.2 |
| BSA Foundries Ltd | Fig. 3.6 |
| BSA Precision Castings Ltd | Fig. 7.4 |
| BSK Aluminium Ltd | Fig. 4.2 |
| Fairey Industrial Ceramics Ltd | Fig. 7.3 |
| Finecast (Maidenhead) Ltd | Fig. 7.4 |
| GKN Squeezeform Division | Fig. 8.12 |
| Glencast Ltd | Fig. 7.9 |
| Holbrook Precision Castings Ltd | Figs. 2.18, 6.4 |
| PI Castings Ltd | Figs. 7.1, 7.2 |
| Precision Products (Cumberland) Ltd | Fig. 6.5 |
| SCRATA | Figs. 4.1, 7.8 |
| Sterling International Technology Ltd | Figs. 5.1, 5.2 |

Mrs V Dhaliwal for typing the manuscript and Miss A M Treasure for typing, revisions and proof reading.

Dr A J Clegg
March 1991

# CHAPTER 1

# Introduction

Metal casting has an historic pedigree as the manufacturing process first used by man to produce intricate metal artifacts and art objects. It played a major role in the industrial revolution and remains the basis of current manufacturing equipment and manufactured goods. The process plays an important part in aerospace component production and, as such, remains at the leading edge of technology development. Despite the competition from plastics and ceramics, metals still remain the dominant materials in the production of capital equipment and manufactured goods. Metal casting will continue to play a major role, as a manufacturing process of considerable versatility, for the foreseeable future.

In common with other manufacturing processes, metal casting has not remained static and there have been significant developments in both the metallurgy of cast alloys and in casting processes. It is the developments in processing, and particularly those devised to improve precision, which are the focus of this book. The objectives of this introductory chapter are to establish the context within which precision casting processes are operated. This requires a definition and description of metal casting, a consideration of the attributes of metal casting in relation to other manufacturing processes, a summary of the structure of the foundry industry, and a definition of the term precision in the context of metal casting.

## Metal Casting

In the metal casting process a metal, or more commonly an alloy, is heated until it is molten, whereupon it is poured into a mould or die that contains a cavity which represents the shape of the component or casting to be produced. In the case of a mould the cavity is produced by a pattern which is a replica of the component required, whereas in the case of a die the cavity replicating the shape of the component is produced by machining. Moulds are invariably expendable, being destroyed when the casting is removed. Dies, on the other hand, are permanent and may be used to produce a succession of castings. Patterns may also be permanent or expendable. A permanent pattern is reusable, whereas an expendable pattern is consumed during the process of mould or casting production. A flow diagram of the metal casting process is shown in Fig. 1.1.

***Sand Casting.*** In the traditional sand casting process, sand is mixed with clay and water to produce a mouldable mixture. The moistened clay provides the bonding agent which binds the sand grains together when the moulding sand is compacted around a pattern to produce a mould. In order that the pattern may be removed from the mould, to provide the cavity into which the metal will be poured, the mould must be made in at least two parts. These mould parts can be separated for removal of the pattern and joined again to receive the metal. The stages in mould production are shown schematically in Fig. 1.2.

***Gating and Feeding.*** The metal enters the mould cavity through a system of channels referred to as the running or gating system. The gating system transports the liquid metal from the pouring ladle to the mould cavity and it has an important influence on the quality of a casting. Metal, like most materials, expands when it is heated and contracts when it cools. Contraction has a major effect on casting quality. During the melting stage the metal is heated to a temperature above its melting point to ensure that it will flow readily and completely fill the mould cavity before solidification commences. When this superheated liquid metal is poured into a mould it will lose heat and contract as it cools, this is referred to as liquid state contraction. With further cooling, and the loss of latent heat of fusion, the atoms of the metal lose energy and become closely bound together in a regular structure. A significant decrease in volume accompanies the change of state from liquid to solid and this is referred to as phase-change or solidification contraction. If no attempt is made to compensate for these two forms of contraction then the casting will show signs of shrinkage, either in the form of surface sinks or as internal voids. To overcome this problem feeder heads are provided and act as reservoirs which feed liquid metal into the casting as it solidifies, thus compensating for the shrinkage. When the entire system is solid it will cool down to ambient temperature and contract as it does so, this is referred to as solid state contraction. As a consequence of this form of contraction the casting will clearly be smaller than the pattern from which it was produced.

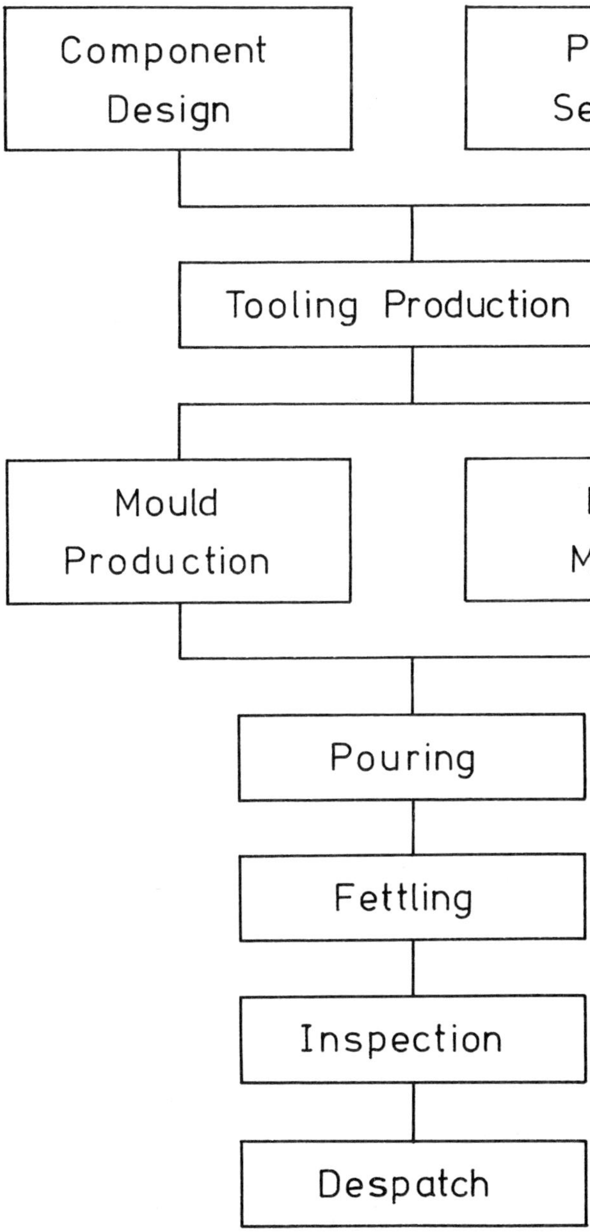


Fig. 1.1 A flow diagram for the casting process

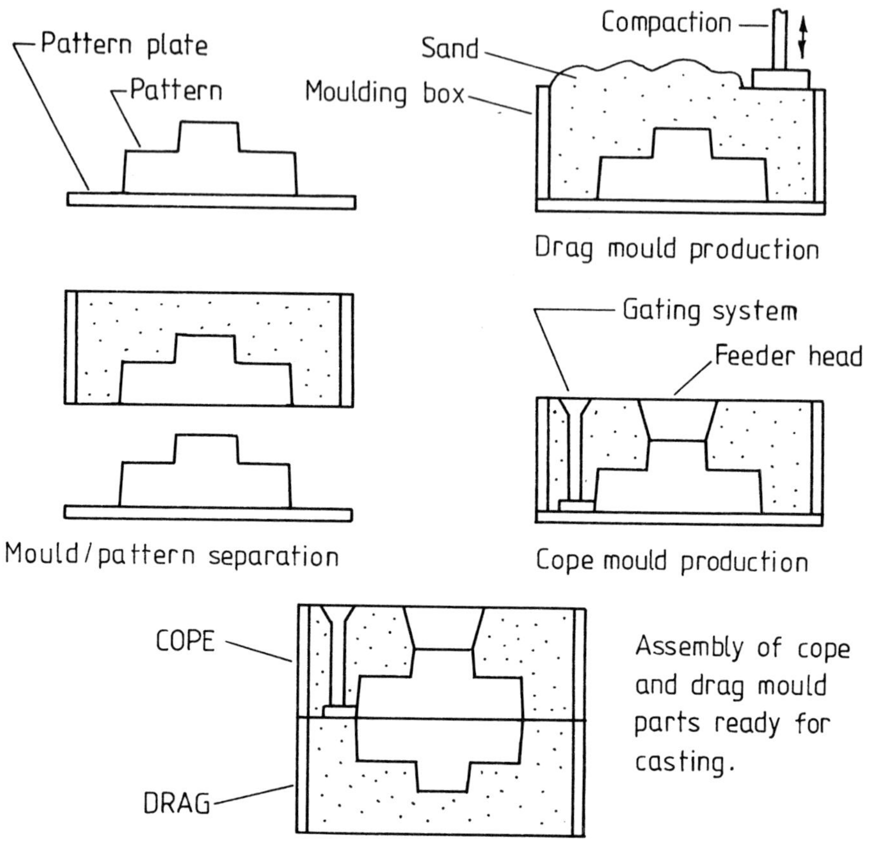


Fig. 1.2 Stages in mould production

***Pattern Features.*** Solid state contraction is, in principle, predictable because it is a function of the metal's co-efficient of thermal expansion (CTE). Consequently, it is possible to produce a pattern which is oversize by the predicted solid state contraction so that the casting produced will be of the required size.

It can be seen that a pattern is not an exact replica of the component to be produced. It is oversize to compensate for solid state contraction and, as can be seen from Fig. 1.2, its surfaces normal to the joint plane are tapered to facilitate its removal from the mould. The pattern will be oversize for a third reason, which is the provision of a machining allowance. This assumes that the casting cannot be produced to the precise dimensions required by the component and that machining will therefore be necessary.

***Expendable Patterns.*** When an expendable pattern is used to produce a mould there is no requirement for a joint in the mould or, consequently, taper on the pattern. The casting produced more closely resembles the final dimensions and shape of the component required. Not surprisingly, therefore, less machining is required and the casting can be considered to be more precise. Expendable patterns are used in the lost-wax, investment casting process and the lost-foam, evaporative pattern casting process.

***Die Requirements.*** As indicated, metal may be poured into a metal die within which the cavity has been machined. Although no pattern is used the die, like an expendable mould, must have a joint line, in this case to enable the casting to be extracted. The die cavity, like a permanent pattern, must incorporate contraction and machining allowances and, to aid extraction of the casting, taper. Castings produced in dies may not be quite as precise as those produced by the lost-wax process, but they are noted for a high degree of precision and consistency.

***Cores.*** Internal features, or indeed external features that cannot readily be moulded, can be provided in castings by the use of cores. Collapsible cores are made in bonded sand, plaster or ceramic materials and are expendable. Separate equipment, termed a core box, is required for the production of cores and special processes have been developed for their production. Cores are located in core prints in the mould or die and these prints are formed as features of the mould or die. Permanent metal cores can be used in the die casting process, especially pressure die casting for which collapsible cores are not currently suitable.

This preliminary consideration of the casting process has provided a basis for process classification according to the permanent or expendable nature of the pattern and/or mould, see Table 1.1.

The list is not exhaustive but does include the principal processes. It is not intended to describe each of these processes, only those which can produce complex shapes will be considered within the context of precision casting.

Table 1.1 *Classification of Casting Processes*

***Permanent Pattern/Expendable Mould Processes***

- Sand Casting Processes
  - Green sand moulding process
  - Dry sand moulding process
  - $CO_2$/Silicate process
  - Shell moulding process
  - Cold set processes
  - Vacuum-sealed moulding process
- Refractory Aggregate Processes
  - Ceramic mould processes
  - Plaster mould process

***Expendable Pattern/Expendable Mould Processes***

- Sand Casting Process
  - Evaporative pattern casting process
- Refractory Aggregate Process
  - Investment casting process

***Permanent Mould Processes***

- Die Casting Processes
  - Gravity die casting
  - Pressure die casting
  - Low-pressure die casting
- Squeeze Casting
- Counter Gravity Casting
- Centrifugal Casting
- Continuous Casting

## The Advantages of the Metal Casting Process

Metal casting competes against several alternative manufacturing processes in the production of shaped metal components, these are:

Forming
Fabrication
Machining
Powder Metallurgy

Each of these processes has its advantages and disadvantages, but none can match the sheer flexibility of metal casting in respect of design versatility, alloy range, shape, size and order quantity.

***Casting Size and Weight Range.*** Components ranging in size from those used to produce a zip fastener, produced as zinc alloy pressure die castings weighing a fraction of a gramme (ounce), to steel mill housings, produced in sand moulded steel and weighing 200 tonnes (197 tons), can be produced by metal casting. No single process meets the entire size and weight range, but there is a process to meet every requirement.

***Design Versatility.*** The foundry maxim "if you can design it, we can cast it" holds true, with the proviso that design modifications should be considered when they can reduce production costs and improve casting quality! However, the designer has almost unrestricted freedom when casting is chosen as the principal manufacturing process. Complex contours, thin sections, ribs, bosses and cast holes can be specified. Internal features such as hollow shapes, volute forms, water jackets and oil galleries can be accommodated through the use of cores. However, the versatility is process specific and an appropriate process must be chosen to provide the principal design features required. As a generalisation, the refractory mould aggregate and sand casting processes provide the greatest versatility and the permanent mould processes the least.

***Alloy Compatibility.*** All metals and alloys in common use pass through a melting and casting stage, even when they are subsequently shaped by forming, fabrication or machining. Even the alloy powders for powder metallurgy can be produced by atomising a liquid melt. Some alloys may only be processed by the casting route, the major examples being the range of cast irons. Whilst metal casting is generically suitable for processing the whole range of commercial metals and alloys, this does not imply that each specific casting process is suitable for every alloy. Constraints may be imposed by the melting temperature of the alloy or by its reactivity with certain mould materials. Nevertheless, by careful selection a process can be chosen to be compatible with a specific alloy system.

***Production Quantities.*** The metal casting process is as suitable for one-off production as it is for one million-off. This versatility is only restricted by the production methods employed by a foundry and the tooling requirements of a particular process. Investment in specialist equipment and tooling usually requires appreciable order quantities to justify it. Sand casting is the most versatile process and is suitable for the economic production of any order quantity.

On the debit side, the metal casting process requires specialist plant and tooling and is energy intensive because of the requirement to melt the metal. Although castings are produced to shape and therefore minimise the amount of material removed by machining, the yield of the casting process itself can be as low as 50%. The waste material, in the form of the gating and feeding system, can be recycled, but there is an additional cost associated with this. Critics of the process will argue that castings are prone to defects such as shrinkage and gas porosity, non-metallic inclusions, distortion and cracking. However, the causes of such defects are well understood and careful attention to process control can eliminate them.

## Structure of the Foundry Industry

The foundry industry has evolved to the extent that individual companies tend to specialise in particular processes, alloys and type of production. This is a general rule to which there are, of course, exceptions. The major division within the industry is based upon whether the companies produce ferrous or non-ferrous alloys. The principal alloys produced according to these headings are listed in Table 1.2. The ferrous sector of the industry is the larger, both in terms of the tonnage output and the number of companies engaged in it. The method of melting and the metallurgy of steels is quite different to that for cast irons, so that most foundries specialise in one of the two ferrous groups. Within the ferrous sector iron founding predominates and this industry features some quite large production units which specialise in the mass production requirements of the automotive industry. The sand moulding processes are predominant within the ferrous foundry sector, with green sand the most widely used mould material, although there is significant and widespread use of the cold set processes. The requirements for making one-off, large castings are quite different from those for making repetition automotive castings, so that companies tend to specialise either as jobbing or repetition founders.

Table 1.2 *The Range of Casting Alloys*

| Alloy Group | Specific Alloys | Principal Casting Processes |
|---|---|---|
| CAST IRONS | Grey cast irons (flake graphite)<br>Nodular irons (spheroidal graphite)<br>Malleable cast irons<br>Heat resistant, chemical resistant and wear resistant alloy irons | sand casting |
| STEELS | Plain carbon steels<br>Low alloy steels<br>High alloy steels<br>Heat resistant, chemical resistant and wear resistant alloy steels | sand casting<br>ceramic mould process<br>investment casting |
| ALUMINIUM | - | sand casting<br>die casting<br>investment casting |
| COPPER | Brass<br>Bronze<br>Gunmetals<br>Aluminium Bronze<br>Silicon Bronze | sand casting<br>die casting |
| MAGNESIUM | - | sand casting<br>die casting |
| ZINC | - | pressure die casting |
| SUPERALLOYS | Nickel-base<br>Cobalt-base | investment casting |
| MISCELLANEOUS | Titanium alloys<br>Bearing metal alloys<br>Magnet alloys<br>Refractory metals | - |

In the non-ferrous sector of the industry the differentiation between jobbing and repetition founders still exists. The size of the foundries in this sector is generally smaller than those in the ferrous sector. Although foundries may specialise in one alloy group, for example aluminium-base, a significant proportion will produce more than one alloy group. The range of processes available to non-ferrous founders is extended beyond those available to the ferrous sector, primarily because they are able to use the die casting processes. Whilst non-ferrous foundries successfully exploit a range of casting processes there is a tendency towards specialisation, especially where pressure die casting is employed. The cost of capital plant and the specialist engineering skills required in die design and production encourages this specialisation.

Investment casting is also highly specialised and represents a distinct sector of the foundry industry. Within this sector of the industry foundries tend to specialise in one of three principal alloy groupings which are superalloys, steels and aluminium alloys. The investment casting industry has shown steady growth for many years and this growth is predicted to continue world-wide. The pressure die casting sector is also growing. In this case the automotive industry is the driving force as it substitutes aluminium alloys for cast iron in its bid to reduce the weight of cars and improve their fuel efficiency.

An indication of the size of the ferrous and non-ferrous sectors of the foundry industry can be obtained by reference to Table 1.3. Not shown in this table are the outputs for the USSR, China, India, Brazil, Korea or Taiwan. It is believed that the output of castings in the USSR exceeds that in the USA and that China would rank third in the World League table.

## Precision in the Context of Metal Casting Processes

If asked to define the attributes they expect of a precision casting, most people would start by stating that the casting must be dimensionally accurate. By this they mean that it should conform as closely as possible to the required dimensions of the final component. Associated with this will be the requirement for consistency, which is an attribute that is especially valued by the customer. Dimensional accuracy is directly related to surface smoothness, so that the attribute of a good surface finish is important in a precision casting. The ability to produce thin sections over comparatively large areas is an attribute much sought after by designers seeking to minimise component weight or achieve consistent heat transfer properties. This is a feature not generally attributed to sand castings but expected in precision castings.

Table 1.3 *Casting Output for 1986 [3-5]*

| Country | Tonnage | | Output | |
|---|---|---|---|---|
| | Ferrous Sector | | Non-ferrous Sector | |
| | Tonnes (x1000) | Tons (x1000) | Tonnes (x1000) | Tons (x1000) |
| Austria | 166.6 | 163.9 | 34.1 | 33.6 |
| Belgium | 194.9 | 191.8 | 14.2 | 14.0 |
| Finland | 90.9 | 89.5 | 6.2 | 6.1 |
| France | 1,798.0 | 1,769.2 | 249.4 | 245.4 |
| Holland | 212.2* | 208.8* | 17.3 | 17.0 |
| Italy | 1,459.5 | 1,436.1 | - | - |
| Japan | 5,563.1 | 5,474.1 | 940.4 | 925.4 |
| Norway | 82.2 | 80.9 | 4.6 | 4.5 |
| Portugal | 78.4 | 77.1 | 6.2 | 6.1 |
| Spain | 703.5 | 692.2 | 88.8 | 87.4 |
| Sweden | 292.0 | 287.3 | 45.0 | 44.3 |
| Switzerland | 196.0* | 192.9* | 19.3 | 19.0 |
| UK | 1,078.0 | 1,060.8 | 142.5 | 140.2 |
| USA | 9,922.3 | 9,763.2 | 3,188.7 | 3,137.7 |
| West Germany | 3,451.5 | 3,396.3 | 587.4 | 578.0 |

* excludes steel castings

The ability to produce intricate shapes and reproduce fine details is expected of precision castings. Whilst the metal casting process is generically capable of producing complex shapes, the replication of fine detail is the province of the specialist precision casting processes. This capability for detail can be extended to the provision of as-cast holes, not necessarily round, which in the best of the precision casting processes may be produced without the need to resort to machining. The final expectation of a precision casting is that it should demonstrate the attribute of metallurgical integrity. The casting should be free from shrinkage and gas porosity, non-metallic inclusions and other casting defects. It should demonstrate, as a consequence, improved mechanical properties and performance in service.

## Factors Influencing the Attainment of Precision in Casting

Several stages are involved in the production of a casting, each of which can exert an influence on precision.

***Tooling.*** A casting cannot be produced to a higher standard of dimensional accuracy or surface finish than the pattern or die from which it is produced. It is essential that the highest standard of tooling in a material appropriate to the process is used. This equipment must be maintained regularly and inspected for damage or wear that might impair the precision of the casting.

***Mould Production.*** In the permanent pattern, expendable mould processes the extent to which the mould conforms to the pattern shape is extremely important. Poor conformity at this stage negates the advantage of accurate tooling. Separation of the mould from the pattern is a critical stage in that damage of the mould is possible. The expendable pattern, expendable mould processes are not susceptible to this problem but the process must still ensure that the mould conforms to the pattern. For the permanent mould processes, mould production implies die production and the requirements here are similar to those highlighted under tooling.

***Mould Materials.*** The material chosen for the mould has a direct influence on both the dimensional accuracy and surface finish of the casting. It may also affect the metallurgical integrity of the casting by influencing the formation of shrinkage, gas porosity, non-metallic inclusions and metal penetration defects.

***Mould Location.*** Processes employing permanent patterns or permanent moulds require a joint line in the mould or die. Joint lines are invariably associated with a reduction in accuracy. Dimensions across a joint line are never maintained to the same tolerances as those within a single mould or die half. This problem is compounded by multiple joints and by the requirements for core location. Further inaccuracies may arise from the need to locate the mould or die halves to one another. The pin and bush method of location is most frequently employed and inaccurate machining or wear of the pins and bushes can contribute to inaccurate matching of the mould or die halves, with a consequent mismatch showing on the casting. Apart from locating to each other mould halves, when produced in moulding boxes, must locate to the pattern equipment, so this aspect must also be considered.

***Pouring Stage.*** The principal requirement at this stage is that the metal should completely fill the mould and die and replicate its shape and detail. This is related to the concept of metal fluidity, which is a technological rather than a physical property. Several factors influence metal fluidity. A high metal pouring temperature promotes fluidity, so too does a preheated mould or die. Alloy composition plays an important role and near-eutectic alloys such as cast iron and certain aluminium-silicon alloys are especially good in this respect. Smooth mould surfaces aid metal flow but if this is achieved at the expense of a loss in permeability, then the back pressure exerted because air cannot readily escape from the mould will in turn restrict the inflow of metal. A second cause of concern is the effect of heat on the mould or die material. This will clearly expand, enlarging the mould cavity and consequently the casting. For those materials which expand in a predictable manner, an allowance can be made in much the same way as the allowance for contraction. However, silica, which is the most common mould material, does not expand in a uniform manner, so that castings produced in this material are the most prone to inaccuracy.

***Casting Contraction.*** Although the value of the CTE is a constant for a given material and is used as the basis for incorporating a contraction allowance in the tooling, casting contraction is unfortunately not so predictable in practice. This lack of predictability arises because castings are complex shapes with section variations and, therefore, do not cool or contract uniformly. In addition, the mould or die may exert a constraint to contraction, building in residual stresses that may lead to distortion. The combination of these factors conspires against accurate prediction of, and allowance for, contraction and therefore contributes to dimensional variations.

***Casting Finishing.*** After separation from the mould or die castings progress through a series of finishing stages (the term fettling is often used) which typically consist of shot blasting, degating, feeder head removal and grinding. Heat treatment may also be required and, in the absence of controlled atmospheres, a separate post heat treatment shot blasting operation may be necessary. The nature of these activities is such that if they are not practised with care they can impair the precision of the castings.

**Measures Available to Improve the Precision of Castings**

Clearly, attention to detail in respect of each of the aforementioned factors will maximise the potential for improving the precision of a casting, independently of the process chosen. However, it is generally recognised that special measures must be taken to optimise precision and some of these measures are introduced in this section.

The sand moulding processes are considered to produce the least precise castings. However, the standards of precision have improved with developments in machinery and processes. In the case of green sand moulding, improvements have been obtained through an understanding of the inter-relationship between clay and moisture content, mould compaction and dimensional accuracy. Moulding machines have been developed which achieve consistently high mould densities and this has resulted in an improvement in dimensional accuracy and consistency. The introduction of the cold setting processes (inorganic and organic) revolutionised the jobbing and short run production of castings. The hardening of the sand in contact with the pattern, a major feature of these processes, improves the conformity of the mould to the pattern and consequently dimensional accuracy and consistency. Hardening of the sand in contact with the pattern is also a feature of the shell moulding process. The process also provides an improved casting surface finish through the use of a fine sand, without the adverse effect on permeability that this usually implies. The development of the unbonded sand processes has also provided improvements. Reduced machining requirements are a feature of the castings produced by both the V-process and lost-foam process. The absence of water or binders from these sands reduces mould material related casting defects and results in improved metallurgical integrity.

Silica sand is the common feature in the sand casting processes and its use legislates against the highest standards of precision in most cases. The exception arises when the mould material is not heated through the $\alpha$ to $\beta$ quartz polymorphic change. Substitution of silica by zircon is practised to a limited extent in the shell moulding process and the Cosworth process, with improved dimensional accuracy as a benefit. However, zircon sand is expensive and this has prevented its widespread use.

The elimination of silica sand as the mould material is a feature of the plaster mould, investment casting, ceramic mould and die casting processes, so that these processes can be expected to demonstrate improved precision. Precision is improved still further in the investment casting process because an expendable pattern is used and there is no mould joint line. With the exception of die casting, these processes also use very fine refractory particles and so provide improved casting surface finishes.

Mould preheat, as an aid to improved metal fluidity and mould or die filling characteristics, is featured in investment casting and is an option in the ceramic mould and plaster mould processes, where it improves thin section capability and detail replication. Die preheat confers the same advantages, but it is the almost instantaneous pressurised filling associated with pressure die casting which makes that process unique in its thin section casting capability. Pressure displacement of metal into moulds or dies is featured in a number of counter gravity casting processes. However, the pressure difference is small by comparison with pressure die casting and its primary objective is to provide quiescent, that is non-turbulent, filling of the cavity. This reduces the generation of non-metallic inclusions, thereby improving metallurgical integrity. The maintenance of the pressure difference during solidification of the casting enhances casting soundness and this is used to advantage in such processes as low pressure die casting, the counter pressure process, the Cosworth process and the Hitchiner CLA and CLV processes.

Vacuum melting and casting is an essential requirement for the production of nickel-base superalloy castings. Under the condition of vacuum, oxidation and gas absorption are eliminated, with a consequent improvement in metallurgical integrity. Vacuum suction may also be used to improve mould filling by removing the air, and thus the back pressure it creates, from the mould or die cavity. In die casting the removal of air from the cavity reduces the occlusion of gases during turbulent filling.

It can be seen from the foregoing comments that there are a number of measures which can be adopted by metal founders to improve the precision attainable in castings. Further details of these measures and other process developments are provided in subsequent chapters.

## Cost Implications of Precision Casting

Precision castings are invariably more expensive than standard castings. This arises because of the more expensive plant and processes required to produce precision castings and through the need for improved standards of process and quality control. The increased cost of precision castings is justified when there is simply no suitable alternative manufacturing route or when the savings which accrue from the reduced requirement for machining outweigh the additional cost of the casting.

Significant cost savings may also occur when precision casting is used to produce components previously produced by an alternative manufacturing process. Regan and Fleck [4] have reported that significant cost savings arose when investment castings were used to replace forgings and/or fabricated assemblies and that this was achieved without a compromise in design function or component integrity. Pressure die castings can provide similar benefits [5] and may even be competitive on a weight saving basis when compared with plastics.

## Near Net Shape Capability

Manufacturing processes that produce components to net or near net shape and size with the minimum number of processing steps and with as little waste as possible have the potential to reduce manufacturing costs and improve manufacturing efficiency. This concept is being applied in metal forming processes such as flashless forging and fine blanking, and in the powder metallurgy route. The metal casting process has an inherent near net shape capability and the precision casting processes can produce castings which come close to meeting net shape and size requirements. With the continuing process development and improved process control, castings are being produced closer to shape and size to the benefit of the industry's customers.

## References

1. Anon. (1989). Foundry statistics. *Modern Casting*, **117**, 1, 22.

2. Anon. (1988). International foundry statistics. *The Foundryman*, **81**, 10, 480-482.

3. Anon. (1989). Japan - a survey of the foundry industry. *Ibid*, 493-494.

4. Regan, T. M. and Fletch, J. N. (1986). Case studies of castings replacing forgings and fabrications in a helicopter engine. *Proceedings of the Advanced Casting Technology Conference*, 103-110. Published by ASM International, 1987.

5. Anon. (1987). Casting saves cash. *Engineering*, **227**, 8, 535-7.

## CHAPTER 2

# Precision Sand Casting Processes (Inorganic Binders)

## INTRODUCTION

Sand is produced by the breakdown of crystalline rocks, principally by the actions of frost and rain. The sand may reside at its place of origin or be distributed by the actions of wind and water. The grain size, shape and distribution will be a function of the history of the sand. For example, sand which remains at its place of origin is likely to be angular and have a wide distribution of grain sizes. Conversely, a sand which has been transported by water is likely to be rounded by the action of grains rubbing against one another and some separation by grain size will have occurred as the finer sand will be carried greater distances by the current. The most commonly used sand in the foundry industry is silica sand, produced by the decomposition of granite. However, there are other sands, such as zircon or olivine, which are used as substitutes for silica. The term sand will be used in its generic sense although, unless otherwise stated, it can be assumed that the use of the word sand implies silica sand.

The properties of a moulding sand are a function of the sand's characteristics and the binder used to bond the sand grains together to form the mould. The sand directly affects such properties as refractoriness, permeability and expansion of the moulding sand. In addition it influences binder demand, moisture demand (in clay-bonded sands) and flowability of the moulding sand, and thereby the very important property of strength. The quality of casting surface finish is a function of sand grain size and distribution. Fine grained sands produce the best surface finish, but at the direct cost of a reduction in permeability and an increase in binder and/or moisture demand.

The purity of a sand is also a very important characteristic as it will directly influence the property of refractoriness and affect its suitability for use with specific binder materials. The presence of sodium and potassium salts or oxides and iron oxide will reduce the refractoriness of the sand, increasing the possibility of melting or fusion of the mould material at the interface with the metal. The presence of alkali material in the sand will neutralise acid catalysts used in the resin bonded cold set processes and an acidic sand will adversely affect bond development in the $CO_2$/Sodium silicate process.

Clay may be present in certain sands and when present in a sufficient amount the mixture may, with the addition of water and appropriate mixing, be used for moulding. Sands of this type are referred to as natural moulding sands. However, the type of clay present in natural moulding sands does not provide the best properties. Consequently, most of the clay-bonded sand which is used in the foundry industry is produced by mixing clay-free sands with specially selected clays. These sands are referred to as synthetic moulding sands.

At this point it is worth reviewing the properties and characteristics of sands in a little more detail.

## The Properties and Characteristics of Sands Used for Moulding

***Refractoriness.*** By refractoriness we mean the ability of the sand to withstand the pouring temperature of the metal without melting. The higher the melting point, the better will be the refractoriness of the sand. Pure silica melts at 1710°C (3110°F), although determining a precise melting point for sands can be difficult as they are unlikely to be chemically pure. For this reason the term softening point is often used. The presence of impurities, as mentioned previously, will impair refractoriness. This is a particular problem in natural moulding sands which contain materials which melt at temperatures as low as 1350°C (2460°F). Such sands may be used with aluminium- and copper-base alloys, and even with cast iron, but they would be totally unacceptable for steels which have pouring temperatures in excess of 1500°C (2730°F). This is another reason for using synthetic moulding sands based on high purity silica sands. By comparison with silica, zircon sand has a melting point of 2420°C (4390°F). Zircon has the chemical composition $ZrO_2.SiO_2$.

***Permeability.*** The property of permeability refers to the ease with which gases will permeate through a bonded sand mould. This property is extremely important when a mould is cast in air. When the metal enters the mould cavity it must displace the air in order to replace it and replicate the mould shape. Failure to do so will result in a short run casting. Some of the air may be displaced through open feeders but the bulk will be displaced through the sand. Large sand grains have large pores between them through which air and any other gases, such as those generated by binder decomposition or water conversion to steam, can readily permeate. Conversely, small grains have small pores which create a more difficult and tortuous route through which the gas must pass. To provide high permeability, therefore, a large grain size would be chosen, but this would be at the expense of poor surface finish. In practice, moulding sands usually consist of a mixture of grain sizes and their distribution may result in infilling of the voids between larger grains by smaller ones, thus reducing the overall permeability.

***Expansion.*** Most materials expand when heated and contract when cooled. Silica sands are no exception to this rule. However, silica undergoes a series of polymorphic changes as it is heated and cooled. At room temperature silica exists as quartz with the $\alpha$ crystal structure. As the $\alpha$ quartz is heated it expands in a regular but comparatively high manner until it reaches a temperature of 573°C (1065°F). At this temperature there is a change from $\alpha$ to $\beta$ quartz which is accompanied by a volumetric expansion, see Fig. 2.1. This can be the cause of expansion-related defects such as scabs but, more importantly, is a major source of dimensional variation in silica sand moulded castings. Other mould materials expand at a lower and more predictable rate and are therefore less susceptible to expansion defects and more likely to produce castings with improved dimensional consistency. A comparison of the expansion characteristics of several moulding materials is provided in Fig. 2.2.

Zircon, as mentioned, has a higher refractoriness than silica and does not combine as readily with iron oxide. It is used in founding steel, as an alternative to silica in sand moulding, in shell moulding, the Cosworth process and for investment casting. It occurs naturally in Australia, India and the USA but is an expensive material for European foundries and must therefore be reclaimed, where appropriate, to minimise moulding costs.

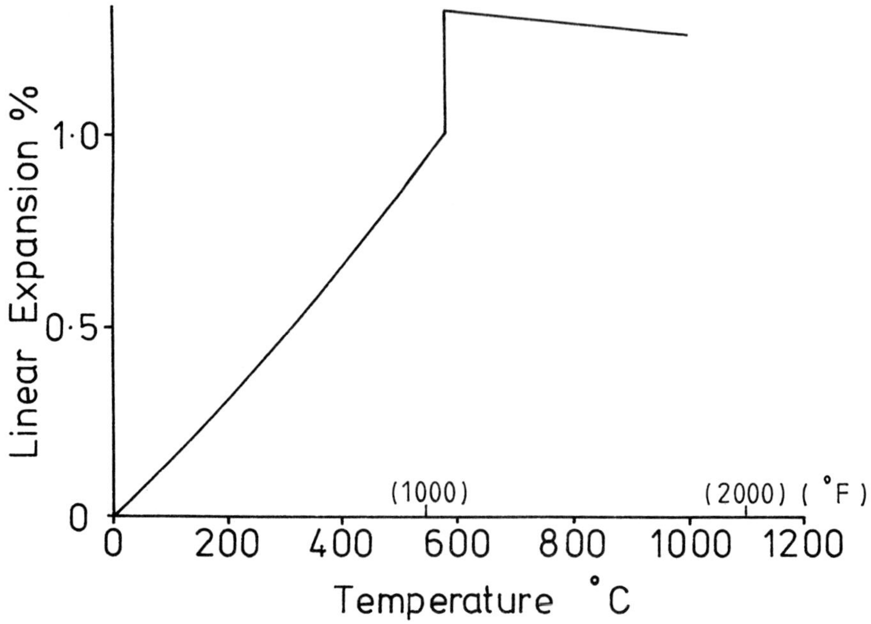


Fig. 2.1 Schematic representation of the expansion of silica

Olivine is a magnesium silicate in which some of the magnesium oxide is replaced by iron oxide, which impairs the material's refractoriness. It is not a widely used material.

Chamotte is an aggregate produced from ground firebricks or burnt fireclay and was used because of its low expansion characteristic for moulds susceptible to expansion defects when silica sand was used.

***Strength.*** Sand by itself has no bonding characteristics, although when vibrated to consolidate it a sand mass can be quite rigid, as evidenced by the unbonded sand processes such as the V-process and the EPC process. In general, however, a binder is used to provide strength in the sand mould. Grain size, shape and distribution, and to a much smaller extent grain surface, influence the strength of the mould in several ways. For a given amount of binder a sand consisting of large grains would provide higher strength than a sand consisting of small grains, because the larger grains would present a smaller surface area and would therefore have a thicker coating of binder. To achieve the same strength the finer sand, with the larger surface area, would require a higher binder addition, that is, it would have a higher binder demand. For a sand in which there is a distribution of sand grain sizes such that smaller grains infill the voids between larger grains, and still smaller grains infill the remaining voids, a still higher strength would be obtained with the given amount of binder. This occurs despite the fact that there is clearly an increase in surface area to be coated by the binder. The increase in strength is explained by the increase in the number of points of contact between grains coated by the binder.

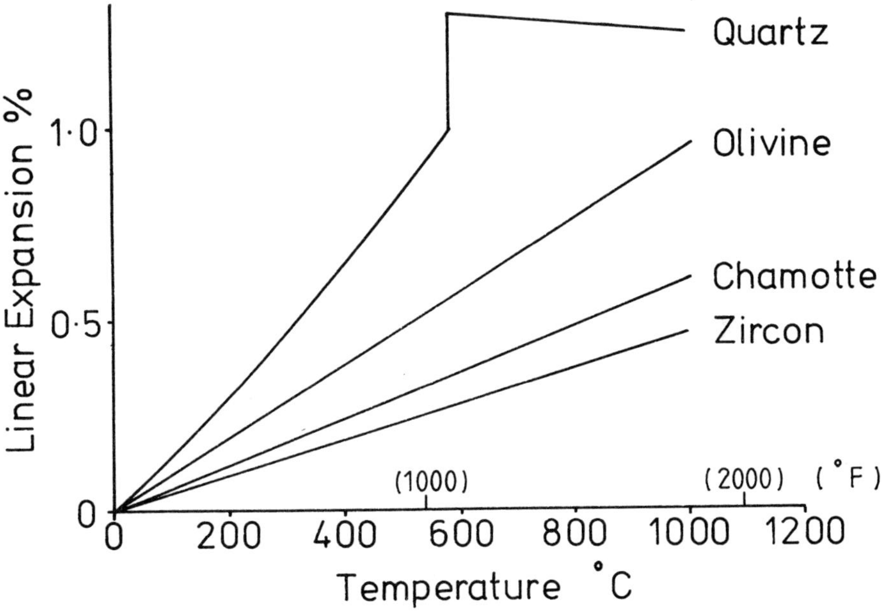


Fig. 2.2 Schematic representation of the expansion of selected mould materials

Grain shape influences strength in two ways. Firstly, in relation to the way that the binder coats the grains and how the individual grains contact one another.

For example, rounded grains have an even coating of binder and present good contact area to their neighbours, but angular grains have uneven coatings and poor areas of contact. Secondly, to achieve optimum strength, moulding sands must be compacted in some way to maximise their density and improve the strength at the points of contact. Consequently, the sands should flow readily and this is a function of shape in that rounded grains flow more readily than angular grains which tend to interlock.

The influence of the grain surface is small by comparison with the preceding factors, but it is likely that grains with a very smooth surface do not provide as good a surface for the binder to key to as a grain with a roughened surface.

The features of grain size, shape and distribution that affect bonding, permeability and surface finish are shown schematically in Fig. 2.3.

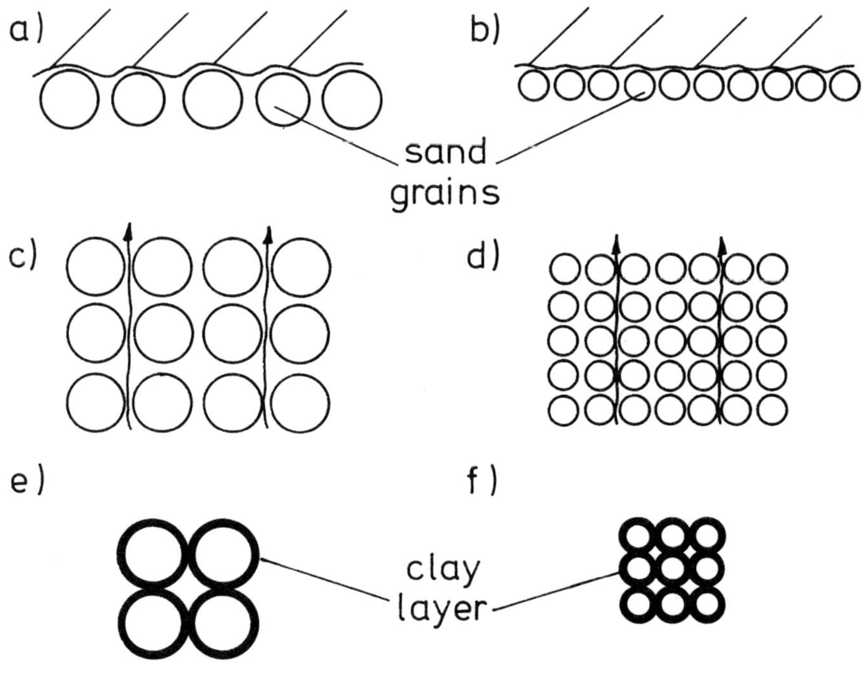

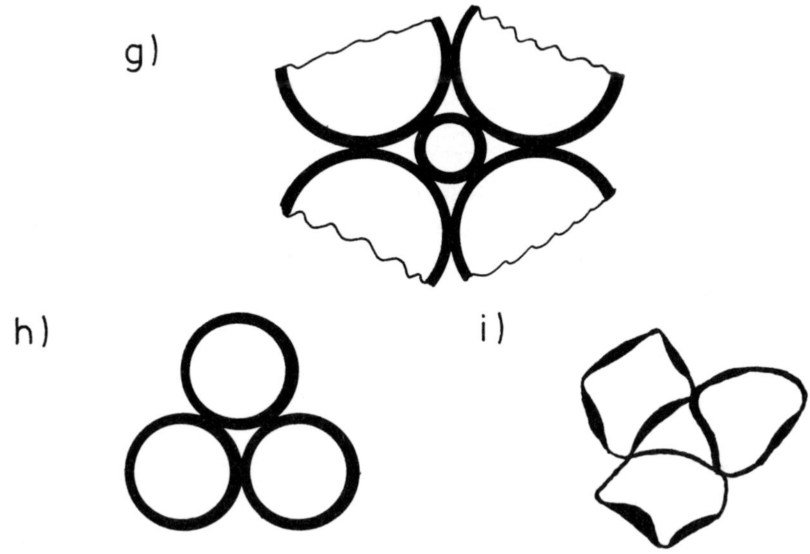


Fig. 2.3 Schematic representation of the influence of grain size, shape and distribution on moulding sand properties: a) and b) the effect of grain size on casting surface finish; c) and d) the effect of grain size on permeability, showing the increased friction associated with the passage of the gas through sands of small grain size; e) and f) the effect of grain size on the thickness of the clay layer; g) additional points of bonding contact provided by infilling of smaller grains between larger ones; h) and i) the effect of grain shape on clay coating and the uneven clay coating associated with angular grains, together with the poor contact area provided by points to points and points to flats.

***Surface Finish.*** This property, like permeability, is directly related to sand grain size and distribution. However, the factors which improve permeability impair surface finish and vice versa. Small grains provide a good surface finish and this is used to advantage in the shell moulding process. Other factors influence surface finish, refractoriness is one and metal pouring conditions such as temperature and velocity are others. High metal pouring temperature and velocity are more likely to encourage penetration of the metal between the sand grains, thus impairing the surface finish. In practice a compromise can be reached by employing a coarse, and therefore permeable, sand and coating it with a refractory wash, containing fine particles, to provide a smooth mould surface.

The term sand is not generally used in conjunction with the ceramic moulding processes of investment casting and the Shaw process, or for plaster moulding. However, the mould material is in a particulate form and in general, therefore, the principles described in the preceding sections can be applied to the materials used in these processes.

## THE GREEN SAND MOULDING PROCESS

Clay-bonded sands have provided the principal medium from which moulds for castings have been produced for centuries. In essence the mould material consists of sand, usually silica in the $\alpha$ quartz form, clay and water. The water develops the bonding characteristics of the clay which binds the sand grains together. Under the application of pressure the mould material can be compacted around a pattern to produce a mould having sufficient rigidity to enable metal to be poured into it to produce a casting. When the mould is used in its moist condition it is referred to as green and the method of producing the moulds as the green sand moulding process. If the mould is dried at a temperature just above 100°C (212°F) the majority of the free moisture will be removed. This is the principle of the dry sand moulding process. Removal of the free moisture is accompanied by a significant increase in the strength and rigidity of the mould. This enables the mould to withstand much greater pressures and so, traditionally, the dry sand process has been used in the manufacture of large, heavy castings.

The relative abundance of sand and clay minerals means that the material costs for the process are low. The mould material is reclaimable, with between 90 and 95% of the sand being recycled, although new sand and additions are required to make up for the deliberately discarded loss. From an environmental view the process is also acceptable, although problems may arise from the use of organic additions to the sand and from the generation of fine silica dust. These features, combined with the relative ease of mould production, have ensured that the green sand moulding process has remained as the principal method by which castings are produced. It was estimated in 1979 that 80% of castings production in the UK was by green sand moulding [1] and in Federal Germany a figure of 67% was quoted for 1986 [2].

The green sand process remains unchallenged as the process capable of the mass production of castings at low cost, especially in cast irons. However, the engineering industry has progressively increased its demand for castings with improved dimensional accuracy, surface finish and soundness. This led to the development of the so called high-pressure moulding machines which produce the high density moulds necessary to meet the customer's demands [3].

**Mould Materials and Properties**

The characteristics of sand that influence mould properties were considered in the introduction to this chapter. The principal addition to the sand is clay which, in the presence of water, provides the bond necessary to develop mould strength. In addition to water and clay, the moulding sand may contain organic materials to improve moulding characteristics and mould properties. In this section the materials added to the sand will be reviewed, together with the method of sand preparation and the development of properties.

The sand used for green sand moulding must fulfil a number of requirements [4]:

(i) It must pack tightly around the pattern, which means that it must have flowability.

(ii) It should be capable of being deformed slightly without cracking, so that the pattern can be withdrawn. In other words, it must exhibit plastic deformation.

(iii) It must have sufficient strength to strip from the pattern and support its own weight without deforming, and to withstand the pressure of molten metal when the mould is cast. It must therefore have green strength.

(iv) It must be permeable, so that gases and steam can escape from the mould during casting.

(v) It must have dry strength, to prevent erosion of the mould surface by liquid metal during pouring as the surface of the mould cavity dries out.

(vi) It must possess refractoriness, to withstand the high temperature involved in pouring without melting or fusing to the casting.

With the exception of refractoriness, all of these requirements are dependent on the amount of active clay present and on the water content of the mixture.

***Clays.*** These materials are hydrous alumino-silicates and consist of very small plate-like particles. The clays of importance to the foundry industry are classified mineralogically as illites, kaolinites and montmorillonites. The latter is the clay of most importance as it is used in synthetic green sand preparation. Montmorillonite has the basic formula $Al_2O_3.4SiO_2.H_2O$. However, some of the aluminium ions (Al+++) are replaced on a one-to-one basis by magnesium ions (Mg++). This creates an excess negative charge which can be balanced by other cations such as sodium (Na+) or calcium (Ca++). The kind of cation has an important influence on the properties of the clay and in defining a montmorillonite the name of the cation is added as a prefix, e.g. sodium or calcium montmorillonite [5]. The principal source of sodium montmorillonite is Western bentonite from deposits in Wyoming and South Dakota in the USA. Calcium montmorillonite sources are more widely distributed and include Southern bentonite from Mississippi, Fuller's earth in the UK and deposits in France, Germany, Greece, Italy and North Africa [5, 6].

The properties of the two bentonite grades are different, sodium bentonite bonded sands have a higher wet strength and are less prone to expansion defects. However, the calcium bentonite develops a green bond more readily and produces higher green strength than the sodium bentonite. The resultant moulding sand possesses better flowability but the mould has a lower dry strength and is prone to erosion and expansion defects. Calcium bentonites may be activated by a base-exchange process in which calcium ions are replaced by sodium ions. This improves the characteristics of the clays for moulding purposes. Blends of sodium and calcium bentonites may also be used to combine the desirable features of each and these blends are particularly suitable for the production of high density green sand moulds [7].

The development of the clay bond depends upon the hydration of the clay. The mechanism of bond formation is described by Sarkar [8] in terms of a net intermiceller force at the sand-clay and clay-clay interfaces which is initiated by the preferential adsorption of cations and anions on hydrated clay surfaces.

***Organic Additions.*** Small additions of starch (about 0.5%) may be added to green sands for high pressure moulding to improve the deformation characteristics. Sands for use in the iron foundry industry may also contain coal dust or a coal dust substitute. These materials are added with the objectives of improving the surface finish and reducing the incidence of sand expansion defects of the castings.

***Sand Mixing.*** Sand preparation is clearly important to ensure that the ingredients are uniformly mixed. Bond development occurs rapidly in synthetic sands milled in high intensity mills [4]. A typical green sand for high pressure moulding would contain between 5 and 7% clay, 2.5 to 3.5% water, 0.5% starch and, when appropriate, up to 3% coal dust. Such a sand would develop a green compression strength around 140 kPa (20 $lbf/in^2$) and have a compactability value between 35 and 45% [4].

***Property Relationships.*** As described previously, there is a direct relationship between grain size and distribution, binder demand and strength development for any refractory aggregate system. In the case of clay-bonded, green sand systems there is also a need to understand the relationship between clay content, moisture content, sand properties and moulding characteristics. Figure 2.4 shows a schematic representation of the relationship between three sand properties: green compression strength, permeability and shatter index (an indicator of plastic deformation), versus moisture content. Figure 2.5 shows the relationship between clay content, moisture content and green compression strength.

For a given clay content strength increases with increasing moisture content to a maximum value, whereupon it begins to fall. Increasing the amount of clay increases the maximum strength attainable but with the requirement for a higher moisture content. The ideal moulding characteristics of a clay-bonded sand are not determined solely by the green strength. To obtain high mould rigidity the green strength must be high, but there must also be some deformation in the sand to enable the pattern to be separated from the mould without the latter breaking. The sand becomes more deformable at higher moisture contents, so that sands are generally used with a moisture content which is just above that required to develop the maximum green strength. Tests for green strength and deformation are complemented by that for compactability, which is a measure of the sand's mouldability.

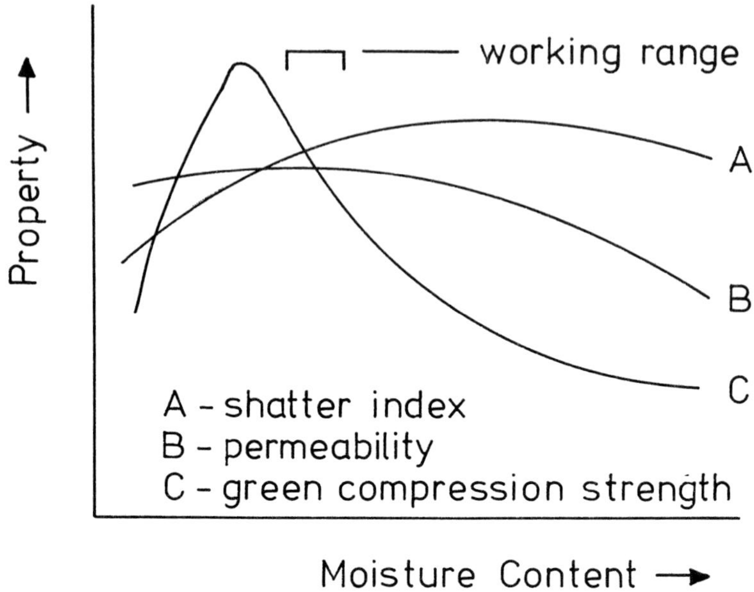


Fig. 2.4 Schematic representation of the relationship between sand properties and moisture content in clay-bonded sands

During the casting process the mould can be heated to quite high temperatures. This may result in the removal of the water of combination associated with the clay. In the case of montmorillonite clays this water is lost over the temperature range 600 to 650°C (1080 to 1170°F). Once this water of combination is lost the clay loses its bonding capability and the clay cannot be reconstituted by adding water, it is therefore referred to as dead clay. Clearly, during each casting cycle, depending upon the temperature reached by the sand, a proportion of the clay will lose its bonding capability. It is not possible to separate this clay from the sand system and, consequently, there will be a build up of fine material in the sand system. Dead clay will reduce the permeability and plastic deformation of the sand but, paradoxically, increase the green compression strength. This latter effect is caused by the infilling effect of the small dead clay particles, which create an increased number of points of contact at which bonding can occur. A small amount of dead clay must be tolerated but the level cannot be allowed to build up too far, otherwise the sand will become brittle and unsuitable for moulding. This explains the requirement for discarding a portion of used sand on a regular basis and its replacement with new sand and additions.

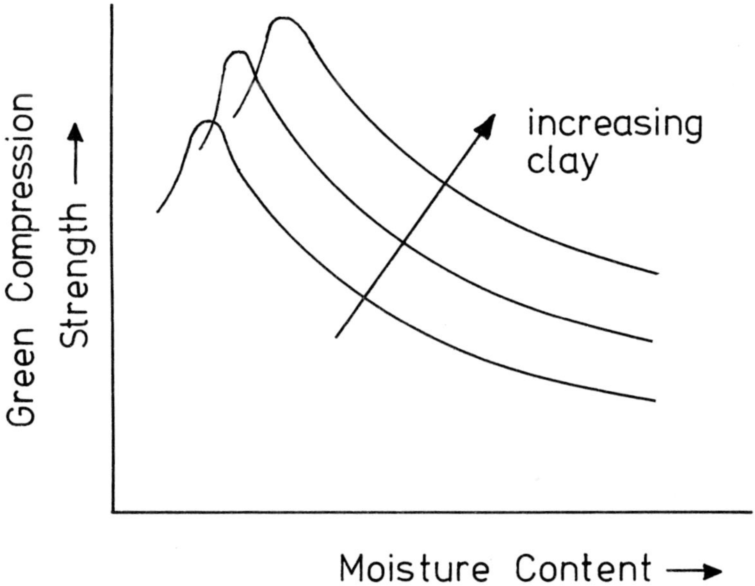


Fig. 2.5 Schematic representation of the effect of increasing clay content on the green compression strength of clay-bonded sand

**Mould Compaction**

The most accurate castings are produced from moulds which accurately replicate the pattern shape and dimensions and resist the forces imposed upon them during casting. A major objective in green sand moulding, therefore, is compaction to attain maximum mould density. However, the mould must display some plastic deformation otherwise it may be damaged during removal of the pattern. Mould hardness can be used as a measure of the degree to which any particular sand has been compacted. The inter-relationship between mould hardness and rigidity, as a function of moisture and clay content and mould hardness and casting dimensional accuracy, is shown in Figs. 2.6 and 2.7 respectively.

A variety of methods can be used to achieve mould compaction, with varying degrees of success. For small quantities, jobbing production moulds are still produced by hand moulding, often using loose patterns. The sand is compacted using hand rammers and the process is slow, labour-intensive and subject to inconsistencies. For larger quantities this approach is unacceptable and machines using the principles of jolting and/or squeezing to achieve compaction of the mould were developed to increase the speed of output and the consistency of the moulds produced. High pressure moulding was introduced during the 1960's to meet the demand for closer tolerance castings and the growth in demand for nodular iron castings [9]. Since the introduction of high pressure green sand moulding there have been a number of moulding machine developments, some of which obviate the need to use squeeze pressure. A brief description of the principal methods of green sand mould production is provided in this section. The machines for mould making with clay-bonded sand can be classified according to the method of compaction, the type of mould and the type of parting [10]. This classification is shown in Table 2.1.

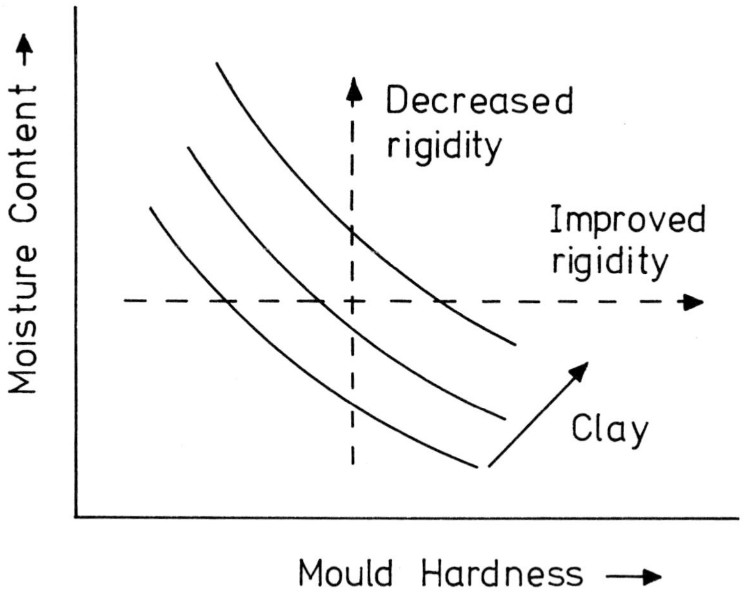


Fig. 2.6 Schematic representation of the relationship between clay content, moisture content and mould hardness for clay-bonded sand

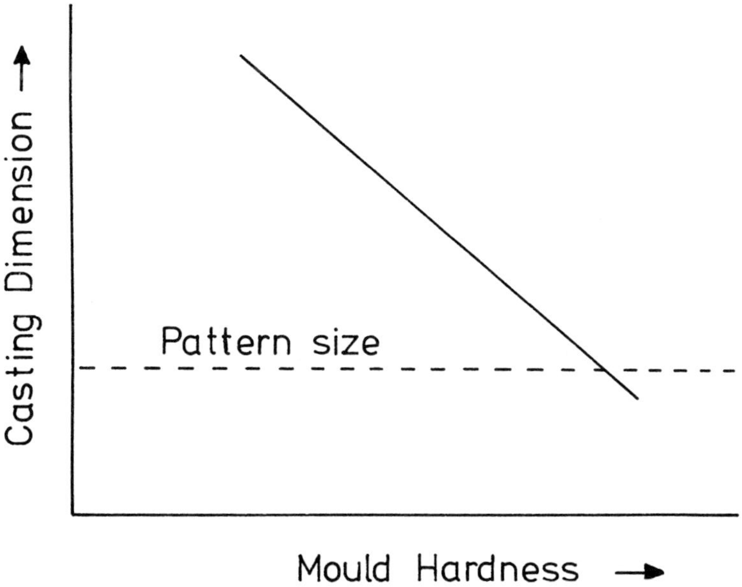


Fig. 2.7 Schematic representation of the relationship between casting dimension and mould hardness for clay-bonded sands

***Jolting.*** A schematic diagram of a jolt machine is shown in Fig. 2.8. The pattern plate, pattern, moulding box and sand sit upon the machine table. The table is lifted under the action of compressed air until the exhaust port is uncovered. Release of the compressed air through the exhaust port reduces the pressure build up and the machine table drops onto the anvil. The sudden deceleration in the downward motion of the sand causes it to be compacted against the face of the pattern and board. The degree of sand compaction will be greatest adjacent to the pattern board and it will be least at the top of the mould. The jolt-only action must be complemented by some other action to overcome this deficiency.

Table 2.1 *Machines for the Production of Moulds with Clay-Bonded Sands [10]*

***Division According to the Compaction Method***

Jolting
Squeezing
Combined Jolting and Squeezing
Shooting/Blowing and Squeezing

Filling the mould chamber by shooting or blowing to achieve precompaction; final compaction by squeezing.

Loose filling of the sand into the mould chamber, followed by precompaction using a short period of air flow; final compaction by squeezing. (Air-flow moulding).

Vacuum-Fill Moulding

Precompaction of the sand by vacuum suction of the sand into the mould chamber; final compaction by squeezing.

Impact (Impulse) Moulding

Loose filling of the sand into the mould chamber, total compaction achieved through the impulse generated by the ignition of a restricted gas-air mixture. (Gas impact compaction).

Loose filling of the sand into the mould chamber, total compaction achieved through the impulse generated by the expansion of a restricted volume of compressed air. (Air impulse compaction).

***Division According to the Type of Mould***

Flask Moulding
Flaskless Moulding

***Division According to the Mould Parting Plane***

Horizontal
Vertical

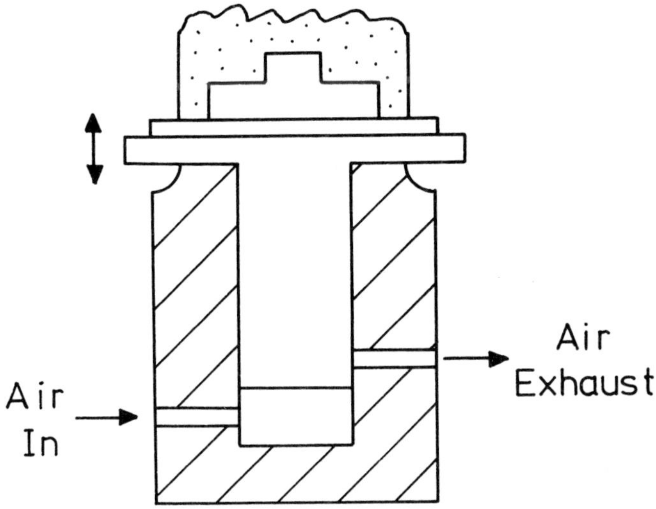


Fig. 2.8 Schematic diagram of a jolt-only moulding machine

***Squeezing.*** In the squeeze-only machine, compaction of the sand is achieved by pressing the sand against a squeeze head, Fig. 2.9. The machine construction is similar to that of the jolting machine but, in order to achieve a high squeeze pressure, a large diameter squeeze cylinder is required. The air pressure is released after a predetermined period and the table is lowered to its rest position. The highest degree of compaction is achieved at the point of contact between the sand and the squeeze board. This method is only suitable for shallow moulds.

***Consecutive Jolt-Squeeze.*** The limitations of the plain jolt and plain squeeze machines can be overcome by a machine design which incorporates both actions. In these machines the jolting action precedes the squeezing action and the machine design effectively incorporates the features shown in Figs. 2.8 and 2.9.

***Simultaneous Jolt-Squeeze.*** An alternative machine design allows jolting to take place during the squeeze operation and by this simultaneous action denser moulds can be produced. Initial jolting, or prejolting, is achieved as described previously and is followed by the squeeze action. At this stage, however, air is again admitted to the cavity under the jolt piston. The piston is constrained from moving upwards because of the squeeze action. However, the piston can move downwards against the action of a spring which it compresses. Release of the air through the exhaust port is accompanied by the release of the compression in the spring and the spring force throws the anvil up against the jolt table thus providing the jolting action.

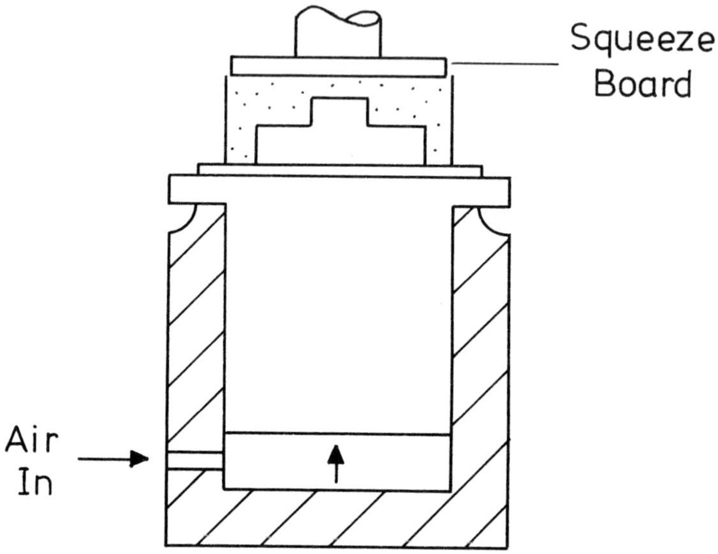


Fig. 2.9 Schematic diagram of a squeeze-only moulding machine

An example of a highly mechanised green sand moulding machine plant is shown in Fig. 2.10.

The machines described so far rely upon the use of compressed air to compact the sand. These machines are limited in their capabilities by the safe working pressure of compressed air lines. This limits the density that can be attained in green sand moulds. To overcome this limitation, hydraulically actuated squeeze heads were introduced as the basis of high pressure squeeze-only machines. Uniformity of compaction was obtained by the use of individually actuated squeeze heads, as featured in the Pneulec Moldmaster design. However, it became clear that even with the higher squeeze pressure obtained by hydraulic actuation, the preliminary distribution of sand within the moulding box is as important as using a high squeeze pressure. During the application of pressure to an uncompacted aggregate a significant proportion of the energy transmitted vertically downwards is lost in the sideways movement of the sand. Hence, some precompaction of the sand will reduce this movement and enable a higher degree of compaction to be obtained in the sand [1].

Fig. 2.10 Photograph of a mechanised green sand moulding plant showing the drag moulds with filters and cores in position

***Shoot (Blow) Squeeze.*** As an alternative to jolting as a method of precompaction, the sand may be blown into the moulding box. Compressed air at a pressure of about 3 bar (45 $lbf/in^2$) fluidises the sand and transports it at a velocity of 15 m/s (45 ft/s) into the moulding box. Compaction is achieved as the velocity of the sand is arrested when it contacts the pattern and board. The pattern and board must be vented to enable atmospheric and compressed air to be discharged. The density achieved by this approach is as high as that by jolting [9] but hydraulic high pressure squeezing is used to achieve final compaction. This approach can be used for horizontally jointed boxed moulds. A similar approach is used in the Disamatic process to produce vertically jointed boxless moulds, see Fig. 2.11.

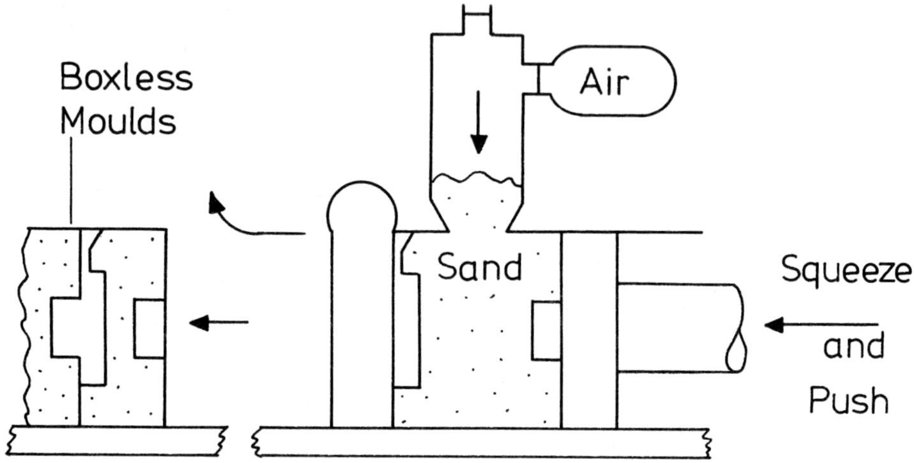


Fig. 2.11 Schematic diagram of a Disamatic moulding machine

***The Disamatic Machine.*** Two half patterns are used, one to form the front and the second the rear of the mould. The pattern to form the rear of the mould is mounted on a hinged door at the front of the machine. The second half pattern is mounted on a plate attached to the squeeze piston. With the door in the closed position sand is blown into the mould chamber. The squeeze piston is then actuated and compresses the sand between the two pattern plates. Following this operation the hinged door retracts horizontally and then swings upwards. The squeeze piston pushes the completed mould out of the machine and closes it against the previously made mould. The string of moulds is advanced one mould's distance along the conveyor track towards the casting station. Moulds without cores can be produced at rates up to 360/hour. Automatic core setting devices can be used with the machine to produce cored moulds at a rate of around 300/hour. A photograph of such a machine is shown in Fig. 2.12.

Fig. 2.12 Photograph of a Disamatic moulding machine showing the core-placing device on the left and the string of boxless moulds emerging from the machine centre-right

***Air-Flow Moulding.*** Developed by the Sintokogio company in Japan during the latter part of the 1970's, this process is also known as the Seitatsu, literally silent moulding, process [4, 11]. The principle of the process is shown schematically in Fig. 2.13. Sand is introduced to the moulding box which is next covered with a hood. Compressed air at a pressure of around 6 bar (85 $lbf/in^2$) is applied to the moulding sand and is exhausted through vents in the pattern and plate. The downward movement of air through the moulding sand at high speed, and the difference in the air pressure between the grains of sand caused by this movement, compacts the sand. This is followed by a squeeze action.

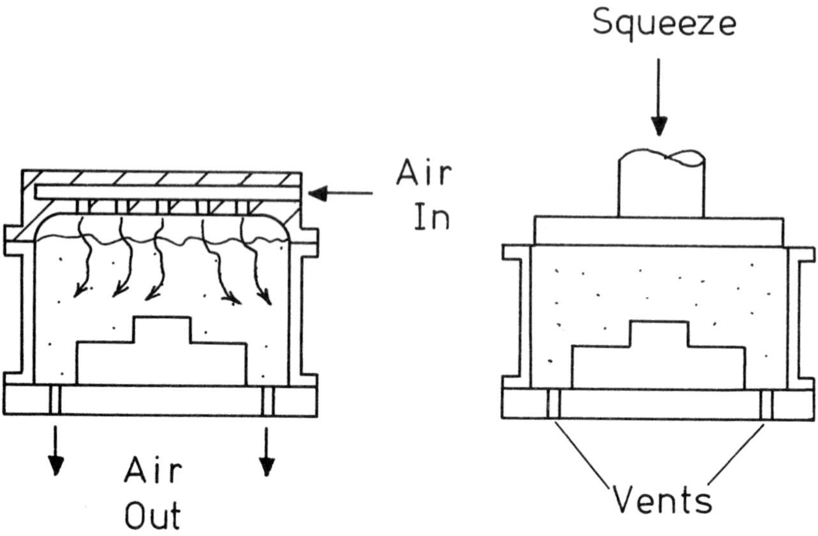


Fig. 2.13 Schematic diagram of the air-flow moulding machine

***Vacuum-Fill Moulding.*** In the Vacupress process, developed by Kunkel Wagner, the sand is drawn into the moulding box by vacuum suction rather than being blown in. An even filling of the mould and an absence of air pockets contribute to a uniformly high mould density achieved by a squeeze action whilst the vacuum is maintained.

***Impact (Impulse) Moulding.*** In this system of moulding, developed by the Swiss company George Fischer, an air pressure or gas pressure wave is used to compress the sand to a suitable level of density without the need to use squeeze pressure.

In the air impulse method, compressed air at a pressure of between 2.5 and 4.5 bar (35 to 65 lbf/in$^2$) is instantaneously released so that the resultant pressure wave acts to compact the sand sitting in the moulding box, see Fig. 2.14. In the gas pressure variant a homogeneous mixture of natural gas and air is ignited and the explosive energy is used to provide the pressure wave which compacts the sand, see Fig. 2.15.

The moulds produced by impact moulding feature a high density layer of sand at the mould/pattern interface, with the density being progressively lower as the distance from the pattern face increases. This is claimed to provide improved mould permeability characteristics [11].

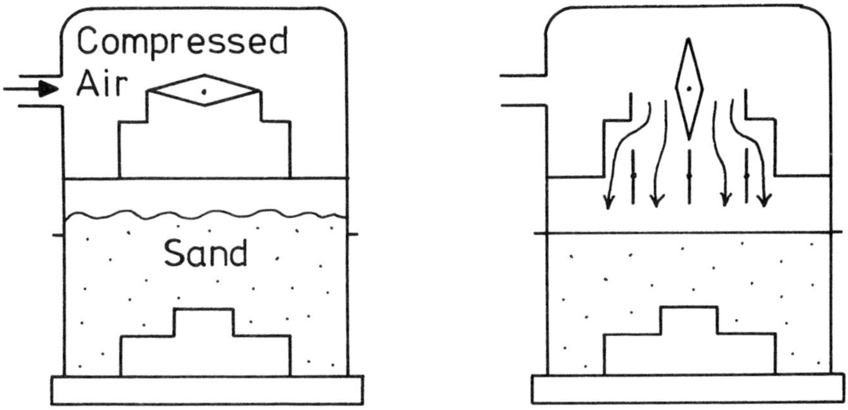


Fig. 2.14 Schematic diagram of the air-impulse moulding machine

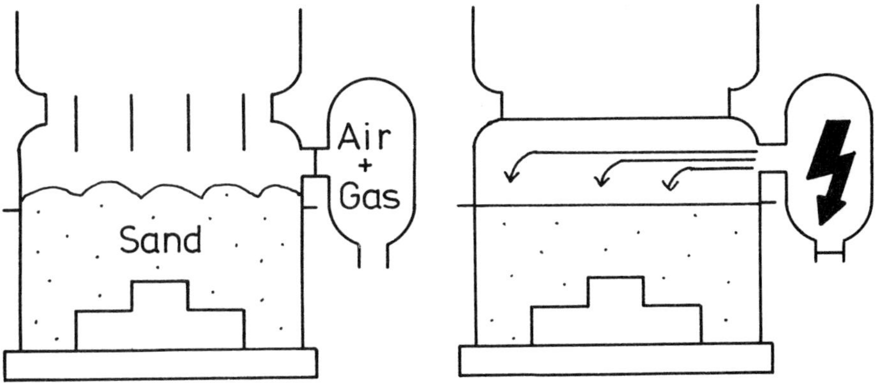


Fig. 2.15 Schematic diagram of the gas-impulse moulding machine

## Tooling Requirements

The two materials most commonly used for pattern construction are wood and aluminium alloy. The former is used for jobbing production and the latter for mass production. Patterns will incorporate allowances for contraction and machining, taper to assist removal from the mould (typically 2 to 5°) and coreprint provision for the location of cores in the mould. The pattern must be manufactured with the knowledge of how the mould will be produced so that an appropriate joint plane can be selected. Tooling for mechanised moulding will incorporate provision for the gating and feeding systems and, where appropriate, mould part location.

### Mould Preparation for Casting

All green sand moulds, whatever the method of production, are jointed and provide access for coring and a need to ensure the accurate location of mould parts. The methods of core setting and mould location will depend upon the moulding system adopted. Highly mechanised moulding production is usually accompanied by automatic pouring, with an associated improvement in metallurgical integrity.

### Gating and Feeding

The normal principles of gating and feeding apply to the production of castings by the green sand moulding process. The design and location of gates and feeders will depend on the joint plane orientation, i.e. whether it is vertical or horizontal.

### Casting Finishing

Following their removal from the mould, castings will be separated from their gating and feeding systems and shot blasted. Gate and feeder pads and joint line flash will be removed by grinding or cutting techniques dependent upon the material.

### Reclamation

The sand is reclaimable, subject to the necessity to avoid a build-up of dead clay in the system.

### Health and Safety Aspects

The process is inherently safe, as the danger of silicosis from fine silica dust is very small as a result of effective extraction systems. The most concern arises from the use of coal dust in iron foundry sands, which generates carcinogens.

This risk is reduced by coal dust substitutes, although these materials may not be as effective as coal dust.

**Process Capabilities**

***Alloy Range.*** The process is suitable for the majority of aluminium- and copper-base alloys, steels and cast irons. By far the greatest output of green sand moulded castings is in cast irons.

***Casting Size and Weight Range.*** The bulk of production is of castings weighing less than 25 kg (55 lb). However, larger castings are readily produced providing that the moulding method achieves the necessary mould rigidity.

***Production Volumes and Lead Times.*** Jobbing production using cheap patterns can result in relatively short lead times. As production requirements increase, better quality, usually metal, patterns are required and the lead time is extended. However, with machines capable of producing moulds at rates in excess of 300/hour, large quantities of castings can be produced relatively quickly.

***Thin Section Capability.*** The green sand process is not renowned for its thin section capability. A section thickness of 5 mm (0.20 in) represents a routinely attainable value.

***Detail Replication.*** The process is not suitable for castings requiring fine detail.

***Surface Finish.*** Generally, castings can be produced with surface roughness values in the range of 6.3 to 25 μm (250 to 1000 μin).

***Cast Holes.*** It would be unusual to attempt to cast holes smaller than $\phi$ 6 mm (0.25 in).

***Dimensional Accuracy.*** Information on this aspect is not readily available. This may be because it is assumed that castings will require machining.

***Typical Castings.*** The range of castings produced by the process is so diverse that choosing typical castings would be difficult. Castings are supplied to the automotive industry, general engineering industry, machine tool industry, valve industry and many others.

## THE $CO_2$/SILICATE PROCESS

The concept of the cold set process, in which a mould or core hardens in contact with the pattern or corebox at ambient temperature to produce an accurate replica, originated with the introduction of the $CO_2$/Silicate process to the foundry industry in the 1950's. According to Gotheridge [12] the art of mixing sand with a liquid sodium silicate based binder, compacting it into a shape and then passing carbon dioxide gas through it, thereby causing it to harden, was first described in a British patent granted to Hargreaves and Paulson in 1889. However, there is little evidence that the process was used to any extent at that time. The modern process has its origin in Eastern Europe, where it was used during the 1940's [13]. Following the granting of a British patent to Petrzela in 1948 the process was adopted in the UK, Western Europe and the USA during the 1950's. Although the process preceded the organic resin bonded cold setting processes, the latter have been used more widely by the industry, despite the advantages that the $CO_2$/Silicate process can provide. The major reasons for this have been the problems of deterioration of moulds and cores during storage and the poor breakdown characteristics associated with the inorganic silicate bond. In addition, the process requires close control of the $CO_2$ gassing stage if optimum properties are to be developed and there is a restriction on the amount of reclaimed sand which can be used in subsequent mixes. Despite the developments which have occurred to ameliorate these problems, the process is still not considered by many foundries to be as attractive as the organic resin binder systems.

The process has a number of advantages which are successfully exploited, and these include [14, 15]:

- the raw materials are readily available and relatively inexpensive.
- the binders are odourless, non toxic, non flammable and water rinsable.
- the raw materials can be handled safely.
- the system does not contain nitrogen, sulphur or phosphorus; so the danger of their pick-up by the molten metal is eliminated.
- little gas is evolved during casting.
- the minimal organic content of the binders eliminates smoke and fume at the pouring and knock out stages.
- the hot plasticity of the sodium silicate bond is beneficial in minimising sand expansion defects and hot tearing problems.
- the binder is relatively insensitive to variations in sand quality.

- the binder is compatible with silica alternatives such as zircon, olivine, chromite and chamotte.
- the binder is compatible with the green sand moulding system; this is important when used in coremaking because the core material will be released into the green sand system at knock out.

The disadvantages of the system include:

- the need for close control over process variables such as temperature, $CO_2$ flow rate and gassing time.
- the strength of hardened material deteriorates in conditions of high relative humidity.
- poor breakdown characteristics can create decoring difficulties under certain conditions.
- residual sodium compounds in reclaimed sand reduce refractoriness.

The process is used in the mass production of cores; as an alternative to green sand, in the production of castings at the top end of that process's weight range; and in the jobbing production of larger castings to a higher standard of accuracy than can be obtained by the clay-bonded, dry sand process.

## Process Outline

Dry sand, silica, or an alternative, is mixed with a viscous sodium silicate binder. This mixture is transferred into the moulding box and compacted by vibration and/or tamping. Carbon dioxide is introduced to the compacted sand by one of a number of methods, whereupon the sodium silicate gels to produce a rigid mould which can then be stripped from the pattern. The moulds are usually coated with a refractory wash, to improve casting surface finish, before assembly prior to pouring.

## Mould Materials

The principal materials used in the production of a mould or core are sand and sodium silicate. However, carbon dioxide is required to develop the bond and other materials may be used to improve the performance of the mould or core.

***Sand.*** A clean, dry silica (or alternative) sand of relatively high purity (>98% silica) is preferred for the process. As the presence of sodium oxide in the binder reduces the refractoriness of the mixed sand, it is preferable that the base sand be relatively free from those impurities which would reduce its refractoriness. Clay should not be present as it will absorb water from the binder and impair the bench life and strength of the mixed sand. If water is present in the sand it will make control of the gassing procedure more difficult. The grain size and distribution of the sand will exert the usual influence on sand properties. As the sand mass will be hardened by the passage of $CO_2$ gas, a reasonable permeability is required.

***Sodium Silicate.*** The binder used in the $CO_2$/Silicate process is a ternary, liquid mixture of silica ($SiO_2$), sodium oxide ($Na_2O$) and water ($H_2O$). The ternary mixture is produced by steam autoclaving a crushed glass, prepared by heating a mixture of crushed or granular silica with soda ash ($Na_2CO_3$) [16]. Foundry grades of sodium silicate are specified by their silica:sodium oxide ratio and by their specific gravity. The ratios in most common use range from 2:1 (low ratio) to 3:1 (high ratio) and the specific gravity is a function of the silicate to water ratio of the mixture.

***Additives for Improved Green Strength.*** The sand and binder mixture has very little green strength. This does not present a problem when the mixture is allowed to set in contact with the pattern. However, if the process is used in the production of detailed moulds, separation of the pattern prior to hardening may be necessary to avoid mould damage. Under these circumstances small additions of clay or dextrin may be added to the sand at the mixing stage. An addition of water will be necessary to complement the clay addition, otherwise the bench life and strength of the sand will be impaired.

***Additives for Improved Casting Surface Finish.*** Iron founders have used coal dust additions in the sand to improve surface finish. However, the use of a refractory wash to coat the moulds or cores is preferable.

***Additives for Improved Sand Breakdown.*** Unlike organic binders, the inorganic sodium silicate does not break down under the action of heat and, consequently, casting knock-out and decoring may be difficult. The presence of clay and/or organic additions improves the breakdown of the sand. Whilst a variety of materials have been used to improve breakdown, sucrose, added directly to the silicate by the supplier, has been the most widely used breakdown agent.

## Sand Mixing and Storage

The principal objective of the sand mixing stage is to obtain a uniform dispersion of the sodium silicate throughout the mixture and a consistent coating of the individual sand grains. This is most easily achieved in a continuous mixer and such mixers can accommodate the use of powdered additions which should precede the silicate addition. As a guide, between 3 and 4% of sodium silicate, based on sand weight, will be required to develop suitable sand properties. The mixed sand has a good bench life providing that it is not left exposed to the atmosphere. Dehydration of the sand's surface will cause it to crust. This can be avoided by covering the sand, preferably with a moistened cover.

## Tooling Requirements

Patterns may be produced in wood, plastic or metal. A good standard of construction is necessary to avoid damage to the pattern during its separation from the hardened mould. The pattern construction must also take account of the method of gassing to be employed. As there may be a slight expansion of the sand during setting, adequate taper must be provided [15]. Any build up of silicate on the pattern can be easily removed by washing with water.

## Mould Production and Gassing Methods

Mould and core production is relatively straightforward as the sand requires only minimum compaction. A combination of light ramming with vibration provides the necessary degree of compaction for moulds, cores may be blown. The methods available for gassing are described below [15, 17] and are shown schematically in Fig. 2.16.

***Probe Gassing.*** The back surface of the mould is spiked and a probe is inserted through which the $CO_2$ gas flows. This method is relatively inefficient.

***Cup Gassing.*** The back surface of the mould is spiked, as with the probe method. However, in this method a sealed cup is placed over the holes. This provides a more efficient distribution of the $CO_2$ gas.

***Hood Gassing.*** In this method, which is suitable for shallow moulds, or those from which the pattern has been stripped prior to gassing, a hood is placed over the mould and the $CO_2$ gas allowed to permeate downwards through it.

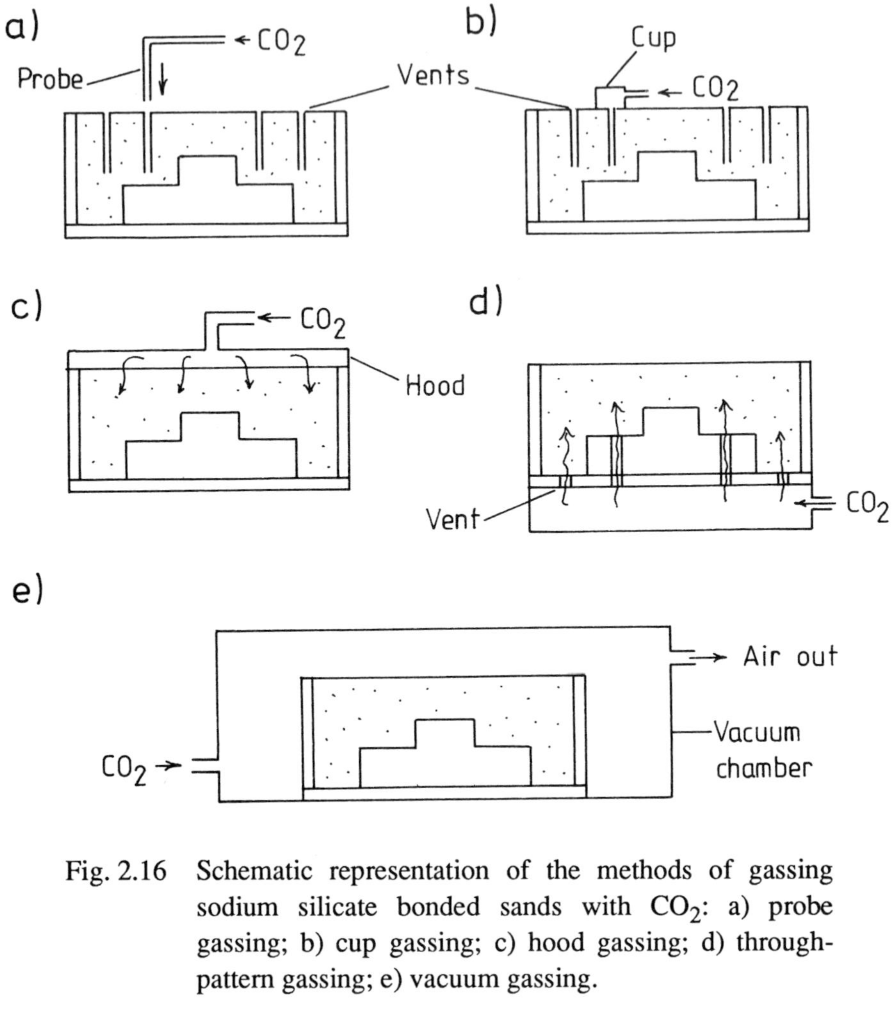


Fig. 2.16 Schematic representation of the methods of gassing sodium silicate bonded sands with $CO_2$: a) probe gassing; b) cup gassing; c) hood gassing; d) through-pattern gassing; e) vacuum gassing.

***Through-Pattern Gassing.*** For deep moulds and those with a large mass, through-pattern gassing provides the most effective gassing method. The pattern/plate assembly is mounted on a plenum chamber from which $CO_2$ gas is supplied through vents located in the plate and pattern.

***Vacuum Gassing.*** The most efficient method of gassing involves placing moulds or cores in a pressure chamber which can be evacuated prior to backfilling with $CO_2$ gas. Hardening is claimed to be even and gas consumption low with this method.

Although, theoretically, the mass of $CO_2$ gas required to harden the sodium silicate is about 11% of the silicate mass [15], the practical requirement depends upon the mass of sand and the time available for gassing. In practice, a realistic usage would be 50% of the silicate mass or 2% of the sand weight, assuming a 4% silicate addition to the sand.

**Hardening Mechanism**

The sodium silicate binders used in the foundry industry are metastable solutions in which the silica is stabilised in solution by a combination of sodium oxide and water [18]. Although there is some conflict in the literature regarding the precise chemical changes which occur during $CO_2$ gassing, it is clear that there are two mechanisms involved.

Firstly, there is a physical mechanism involving dehydration. The passage of dry $CO_2$ gas through the sand mixture removes water and consequently increases the viscosity of the binder which acts as an adhesive. This effect could be achieved simply by blowing dry air through the sand.

Secondly, there is a chemical reaction which removes the sodium oxide from the silicate and precipitates a silica gel. This reaction is promoted by the formation of carbonic acid when $CO_2$ dissolves in water. For a detailed consideration of the chemical reactions which occur, the reader is referred to the paper by Srinagesh [16]. The reaction can be summarised as follows:

$$Na_2SiO_3 + 2H_2O + CO_2 \rightarrow Si(OH)_4 + Na_2CO_3$$

The monosilicic acid ($Si(OH)_4$) monomer is unstable by itself and so combines with similar monomers in what is effectively a condensation polymerisation reaction. During the course of this reaction the majority of the water is lost and the gel-like silica material, which provides the bonding, is produced.

It is possible to overgas the silicate and this is believed to result in the formation of sodium bicarbonate ($NaHCO_3$) rather than sodium carbonate ($Na_2CO_3$). This is accompanied by complete dehydration of the gel, which becomes brittle, so that the sand becomes weaker and more friable [19].

It has been suggested that between 30 and 40% of the ultimate strength is developed as a result of the physical mechanism, with the balance the result of the chemical reaction.

## Process Control

Although the process may appear relatively straightforward to use, careful control must be exercised if optimum performance is to be obtained. The most important aspects to consider are the choice of silicate ratio, the influence of temperature, and the gassing conditions of time and flow rate.

***Silicate Ratio.*** The performance characteristics of low and high ratio silicates are sufficiently different to ensure that a choice must be based on a consideration of the operating conditions, production methods and environment in the foundry. Martin and Ennis [15] have summarised the differences between the silicates as follows:

| | **low ratio** | **high ratio** |
|---|---|---|
| amount of addition | equivalent | equivalent |
| amount of $CO_2$ required | more | less |
| time to harden | more | less |
| initial strength | less | more |
| strength on standing | more | less |
| breakdown | worse | better |
| cost of silicate | more | less |
| tolerance to unskilled handling | more | less |

If the process is to be applied to the mass production of cores which are to be used promptly, then the choice would be a high ratio silicate. Conversely, if the process was being used for the jobbing production of moulds which might not be cast for a day or two, then the choice would be a low ratio silicate.

***Temperature.*** The temperature of the whole system should ideally be maintained between 16 and 32°C (60 and 90°F) [12]. Below 10°C (50°F) the silicate becomes increasingly viscous and there is the danger that it will not be evenly distributed during mixing. The reaction rate between the silicate and $CO_2$ will be slowed down, an effect which is exacerbated by the endothermic nature of the physical dehydration action which is not compensated by the slightly exothermic nature of the chemical reaction. At temperatures above 32°C (90°F) premature dehydration of the silicate is likely and this may lead to loss of control, especially in regard to strength development. Facilities for the monitoring and control of temperature are therefore essential.

***Gassing Conditions.*** Given the guideline that the amount of $CO_2$ required will be 2% of the weight of sand to be hardened, the flow rate and time for gassing can be determined. However, in practice this approach is too simplistic. Flow rate exerts a major influence on the hardening process. At high flow rates the physical action predominates, whereas at low flow rates the chemical reaction is predominant. Higher as-gassed strengths are developed using low flow rates. However, subsequent strength on standing for 24 hours may be greater when a high flow rate is employed. This relationship exists for low ratio silicates. However, strength tends to deteriorate on standing for high ratio silicates. The time of gassing is also important. Although strength increases progressively as the gassing time is extended there is, in the case of low ratio silicates, an initial period of several seconds at the start of gassing when virtually no strength is developed [17]. The higher ratio silicates offer faster rates of hardening with lower rates of $CO_2$ consumption and are, therefore, particularly suited to mass production operations.

## Storage Characteristics

As indicated in the preceding sections, the strength of hardened sands may change with time during storage. For low ratio silicates, continued dehydration leads to an increase in strength. However, in the case of the high ratio silicates, there is an embrittlement of the bond associated with continued dehydration with consequent deterioration in strength. In conditions of high humidity, water loss is reduced during storage. This arrests the deterioration in strength associated with high ratio silicates.

If proprietary silicates containing sugar-base breakdown agents are used, rather than the straight silicates considered so far, then the storage characteristics are changed. The hygroscopic nature of these additions results in water absorption and a reduction in strength during storage.

**Breakdown Characteristics**

The strength of a mould or core will at first increase as a result of the heat from the molten metal. This increase is caused by the dehydration of the silicate and will reach a maximum at a temperature between 300 and 400°C (570 and 750°F). Thereafter the strength will decrease rapidly as the bond softens. The plastic deformation of the sand is high at this time, which reduces the tendency for expansion defects or hot tears to occur. As the sand cools down it will regain its strength and indeed may display a significant increase in strength, depending upon the temperature reached. The highest retained strength is produced when the sand has been heated to around 900°C (1650°F), see Fig. 2.17 [17].

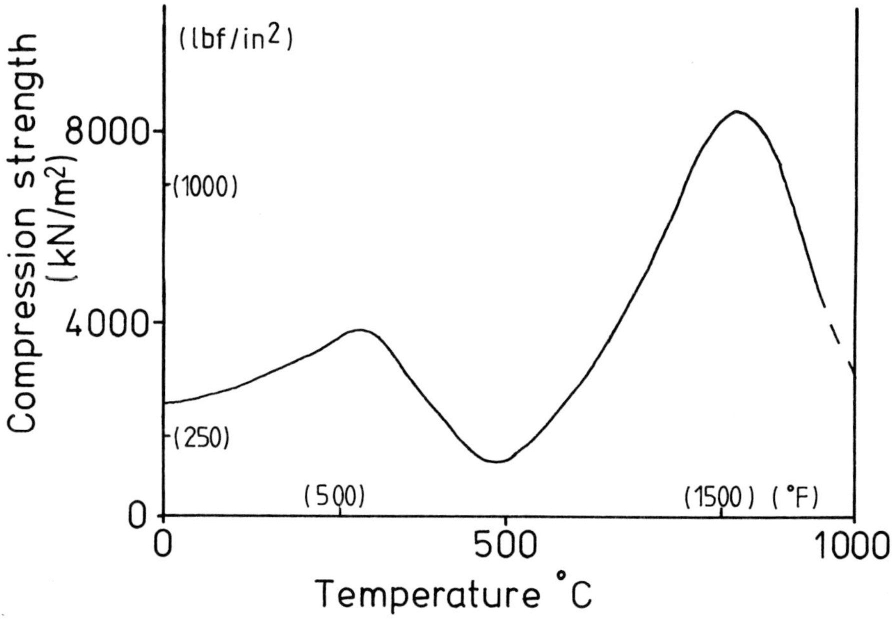


Fig. 2.17 Schematic diagram showing the effect of temperature on the retained (cold) strength of a sodium silicate bonded sand

This high strength makes it particularly difficult to break the sand down at the fettling stage and decoring becomes especially difficult. The problem can be reduced by using high ratio silicates in preference to low ratio silicates and by using the minimum silicate addition commensurate with obtaining the required strength. Additions to improve breakdown are most commonly based on sugar which, being organic, burns out of the sand producing a weaker structure that breaks down more readily. Certain mineral additions are effective because they displace the peak strength to a region of higher temperature [17].

**Sand Reclamation**

The presence of sodium compounds reduces the refractoriness of silica sands, leading to poor casting surface finish. Residual sodium compounds can be removed by treating the sand with water, but the plant is expensive and the water must be treated before it can be discharged through municipal drains. Dry attrition reclamation will not remove all the residual sodium compounds, so that reclaimed sand additions are restricted to about 70% of a subsequent sand mix. This is significantly lower than for organic resin cold set systems.

**Process Developments**

Attempts have been, and continue to be, made to overcome the problems associated with the process, namely the gassing procedure, core and mould storage, and sand breakdown.

Control of the gassing procedure is critical to the success of the $CO_2$/Silicate process and the elimination of the need for gassing would simplify the process. The need for gassing was seen as a disadvantage, especially when the process was compared with organic resin cold set processes. A number of non-gassing methods have been developed for hardening sodium silicate. Those based on powder additions such as ferro-silicon or dicalcium silicate were short-lived for a variety of reasons [14]. However, the method based on a liquid ester hardener has proved more popular.

***The Ester-Silicate Process.*** In this process a liquid ester, constituting 10 to 14% by weight of the silicate, is added to the sand prior to adding the high ratio silicate. Hardening occurs in the following manner [19]. Hydrolysis of the ester in the presence of the alkaline sodium silicate solution results in the formation of a mild acid and an alcohol. The acid reacts with the sodium oxide, which precipitates the silica to provide the bond. Many organic esters can be used to harden sodium silicate but the ones most commonly used are:

| | |
|---|---|
| glycerol diacetate (diacetin) | fast cure |
| ethylene glycol diacetate (EGDA) | medium cure |
| glycerol triacetate (triacetin) | slow cure |

These esters can be blended to give a setting time to suit a foundry's production requirements [20]. The factors affecting the curing of sodium silicate by esters have been examined by Piotrowski and Svensson [21].

***Co-Binder Technology.*** During the 1970's proprietary binders were introduced which incorporated special additives to improve the rate and level of strength development, reduce $CO_2$ consumption, improve storage characteristics and improve breakdown [18, 19]. Starch hydrolysate or alkali soluble phenolic co-binders are non-hygroscopic and so provide improved storage characteristics, as well as contributing to the strength of the sand whilst simultaneously improving breakdown.

### Mould Preparation for Casting

After hardening and separation from the pattern, moulds can be coated with a refractory wash and cores can be positioned if required. Moulds are closed, using appropriate location devices, and are made ready for pouring. Figure 2.18 shows the production of boxless moulds by the process.

### Gating and Feeding

The normal gating and feeding principles apply to the production of castings by the $CO_2$/Silicate process. The high mould rigidity provided by the process may reduce the requirement for feeding in certain grades of cast iron.

### Casting Finishing

Finishing procedures are similar to those described for green sand moulded castings.

### Reclamation

The dry attrition process is recommended for silicate bonded sands. It is recommended that reclaimed sand be restricted to 70% of subsequent sand mixes, although higher levels are claimed for certain proprietary grades of silicate.

Fig. 2.18 Photograph showing the production of boxless moulds by the $CO_2$/Silicate process

**Health and Safety Aspects**

Although accepted as a safe and environmentally acceptable process, due care should be exercised in handling the chemicals associated with the process. Adequate extraction and ventilation is necessary to prevent the build up of $CO_2$ in gassing areas.

**Process Capabilities**

***Alloy Range.*** The process is suitable for a wide range of alloys, including aluminium- and copper-base alloys, cast irons and steels.

***Casting Size and Weight Ranges.*** The process is used for heavy sectioned castings at the top end of the green sand range, around 25 to 50 kg (55 to

110 lb), and for the production of much larger castings as an alternative to the dry sand moulding process.

***Production Volumes and Lead Times.*** The process is suitable for one-off or medium run production. Lead times are no longer than those for green sand moulding. The production rate is slower than that for green sand but much faster than the dry sand alternative.

***Thin Section Capability.*** As a sand casting process, a section of 5 mm (0.20 in) would be an acceptable minimum value.

***Detail Replication.*** As a sand casting process, it is not renowned for detail replication.

***Surface Finish.*** A surface roughness value in the range of 6.3 to 25 μm (250 to 1000 μin) can be expected.

***Cast Holes.*** In some circumstances it may be possible to cast holes as small as ϕ 6 mm (0.25 in).

***Dimensional Accuracy.*** Whilst the accuracy of castings is improved by the use of the process, machining will still be required.

***Typical Castings.*** Castings for a wide range of engineering applications are produced by the process. These include machine tool castings, valve castings, and large diesel engine blocks and heads.

## THE ALUMINA-PHOSPHATE SYSTEM

This inorganic binder system is composed of two parts: an acidic, water-soluble, alumina-phosphate liquid resin binder; and a free-flowing, powdered metal oxide hardener [20]. The hardener powder is dry-blended with the sand before adding the binder. The coated sand is then compacted in the mould or core box and hardening is allowed to progress. The process is environmentally acceptable and, unlike the silicate processes, has excellent breakdown. Although suitable for a range of alloys, including ferrous materials, the process is not widely used.

Two other inorganically bonded systems which are used to a limited extent by the industry are the cement bonded sand process and the fluid sand process. Further information on these processes can be obtained from the literature [22].

## References


1. Williams, R. (1979). The present status and future place of greensand in the foundry industry. *The British Foundryman*, **72**, 6, special supplement, 53-62.

2. Weiss, R. and Kleinheyer, U. (1987). Current trends in the use of moulding and coremaking processes. *Casting Plant + Technology,* **3**, 4, 10-19.

3. Golka, H. R. (1980). Production of high density moulds by jolt and squeeze. *Foundry Trade Journal*, **148**, 3187, 739-760.

4. *Greensand Technology*. Published by BCIRA, Alvechurch, 1985.

5. Parkes, W. B. (1971). *Clay Bonded Foundry Sands.* Published by Applied Science Publishers Ltd, London.

6. Joseph, R. G. (1980). Understanding green sand: a review. *Modern Casting,* **70**, 1, 64-67.

7. Wootton, R. (1986). Good sand management for best results when using high density moulding. *Foundry Trade Journal,* **160**, 3324, 278-284.

8. Sarkar, A. D. (1967). *Mould and Core Material for the Steel Foundry.* Published by Pergamon Press, Oxford.

9. James, K. T. (1984). Selecting a moulding machine - the foundryman's dilemma. *Foundry Trade Journal,* **156**, 3284, 357-367, 375.

10. Engels, G. and Schneider, G. (1987). New developments in compaction methods and machines for green sand moulding shops. *Casting Plant + Technology,* **3**, 1, 2-13.

11. Schaarschmidt, E. (1985). Modern green sand moulding systems. *Ibid*, **1**, 4, 12-23.

12. Gotheridge, J. (1980). A look into the future: wider application of the sodium silicate/carbon dioxide process through a better understanding of the basic principles and the new technology. *Trans. AFS,* **87**, 669-706.

13. Sarker, A. D. (1964). *Foundry Core and Mould Making by the Carbon Dioxide Process.* Published by Pergamon Press, Oxford.

14. Stanbridge, R. P. and Williams, R. (1980). Recent advances in the technology and application of silicate bonded sand. *Trans. AFS,* **87,** 77-90.

15. Martin, G. J. and Ennis, C. S. (1986). The $CO_2$/Silicate process - the current state of the art. *The British Foundryman*, **79**, 8, 376-380.

16. Srinagesh, K. (1979). Chemistry of sodium silicate as a binder. *International Cast Metals Journal*, **4**, 1, 50-63.

17. Nicholas, K. E. L. (1972). *The $CO_2$/Silicate Process.* Published by BCIRA, Alvechurch.

18. Atterton, D. J. and Stevenson, J. V. (1981). The carbon dioxide process: a look back - a look forward. *Trans. AFS*, **72**, 55-64.

19. Penko, T. (1989). Environment right for silicatc's rediscovery. *Modern Casting*, **79**, 3, 23-26.

20. Carey, P. R. and Winters, D. L. (1986). Updating resin binder processes - part V. *Foundry*, **114**, 6, 46-52.

21. Piotrowski, K. J. and Svensson, I. L. (1985). On the ester curing of sodium silicate based binders. *The British Foundryman*, **78**, 3, 117-126.

22. *Chemically Bonded Cores and Molds.* Published by the American Foundrymen's Society, Des Plaines, Illinois, USA, 1987.

CHAPTER 3

# Precision Sand Casting Processes (Organic Binders)

## THE COLD SET AND COLD BOX PROCESSES

### Introduction

In the production of moulds and cores the highest level of accuracy is obtained when the moulding material sets in contact with the pattern or core box. The first moulding processes to achieve this objective were the inorganically bonded $CO_2$/Silicate process, described in the preceding chapter, and the organically bonded shell moulding process. The former is the original cold box process, the latter requires the application of heat to cure the mould. Like the $CO_2$/Silicate process, the shell process can be used for producing cores. The need to apply heat to cure the bond is time consuming and the hot setting processes are associated with unpleasant fume generation. The development during the 1950's of organic binder systems which could be cured at room temperature revolutionised production in the sand casting foundry industry. The processes which were developed, and continue to be developed, are known by colloquial names such as: air-set, self-set or no-bake; or by trade names which include Alphaset, Betaset, Isocure and Pepset. Some of these processes are ideal for the production of large moulds for one-off castings, others for the series production of medium size castings, and yet others for the mass production of cores. The adoption of these processes has enabled the foundry industry to improve productivity and quality. In comparison with the processes used traditionally by the sand casting industry, these processes can be considered to be precision sand casting processes. In this chapter the organic binder processes will be defined and two specific precision sand casting processes, namely the Cosworth process and the shell moulding process, will be described in more detail.

In the opening section, which defines the organic binder systems that can be cured at room temperature, two groups of processes will be considered: cold set processes and cold box processes. The term cold set will be used to describe those processes in which the sand, binder and curing agent are mixed together at the outset, so that curing commences immediately, with the cure time depending upon the type and amount of curing agent. The term cold box will be used to describe those processes in which the sand and binder are mixed together and shaped into a mould or core before the curing agent is introduced in the form of a gas or vapour. These processes are characterised by rapid cure times.

### Cold Set Processes

***Furan Binders.*** The term furan is a generic one which is used to classify those binders based on the use of furfuryl alcohol by itself, or modified with urea, phenol, formaldehyde or other additions. These binders are cured at ambient temperature by the addition of an acid catalyst which produces a cross-linking action to develop the bond. The furan binders were introduced to the foundry industry in the late 1950's.

The binders containing the highest quantity of furfuryl alcohol provide the best properties but are the most expensive. These binders would contain the lowest water and nitrogen contents and be sold as premium grades for the most demanding quality applications. As the amount of nitrogen (introduced in direct proportion to the amount of urea in the resin) and water (present as a result of the method of resin preparation) increases, the cost of the binder decreases. However, the application of these binders is limited because the nitrogen content may cause nitrogen pin-hole defects in ferrous castings, and the water content may result in hydrogen gas defects in non-ferrous castings. Silane is generally added to the binders to act as a coupling agent, improve tensile strength and improve the bond's resistance to degradation by moisture.

The choice of catalyst depends upon such factors as the reactivity of the binder, the rate of cure required and the type of metal to be cast. Binders containing urea are relatively reactive and readily catalysed by phosphoric acid. For the low-nitrogen binders the more reactive aryl sulphonic acids, such as benzene-sulphonic or toluene-sulphonic, may be used. Sulphuric acid can be used to improve the reactivity of catalysts, but its presence generates sulphurous fumes on casting and there is a danger of sulphur pick-up by SG irons.

The rate of cure will be accelerated by higher ambient or sand temperatures and decelerated by the presence of moisture in the sand or by high relative humidity. The generation of water during curing by the condensation polymerisation reaction can reduce the degree of through cure if it is unable to evaporate.

The binder addition is typically of the order of 0.8 to 2.0% of the sand weight and the catalyst addition between 20 and 60% of the binder weight. The furan binders provide excellent dimensional accuracy, a high resistance to sand/metal interface casting defects, yet decompose readily after casting to provide superior breakdown [1].

***Phenolic Binders.*** Although introduced at the same time as the furan binders, they were comparatively slow to cure, and it was not until the 1970's that developments in resin and catalyst chemistry, and the shortage of furfuryl alcohol, combined to increase their popularity as an alternative to furan binders [1]. Resole phenolic resins, which contain phenol, formaldehyde and modifying additions in water, are the basis of these binders. Despite the presence of between 10 and 25% water, the cure rate is not adversely affected. Phosphoric acid will not provide a sufficiently fast rate of cure, so that a sulphonic acid catalyst is preferred and is added at a level of between 20 to 45% of the weight of the binder.

***Ester Cured Phenolic Binders.*** These binders were introduced to the foundry industry in the early 1980's under the trade name Alphaset and were quickly adopted by the steel foundry industry [2]. The phenolic resole resin binder contains no furfuryl alcohol or nitrogen and does not require a sulphur-containing catalyst to harden it. The two-part binder consists of the water-soluble, alkaline phenolic resin, plus between 15 and 30% of the liquid ester co-reactants. A blend of esters is chosen to provide the required setting time but cause only partial cross-linking of the phenolic resin [2]. As a result, the bond retains a small degree of thermoplasticity which enables the sand to absorb some expansion and this, in turn, reduces the incidence of expansion related defects. Full strength is developed in the binder by the application of heat during casting. Figure 3.1 shows a carbon steel casting produced by the Alphaset process.

Fig. 3.1 Photograph showing a carbon steel casting produced by the Borden Alphaset process

***Alkyd Urethane Binders.*** These three-component binder systems are particularly suitable for the production of large moulds or cores requiring extended setting times [3]. The three components of the binder system are:

Part A: an alkyd oil resin, produced by modifying a natural or vegetable oil
Part B: a liquid amine/metallic catalyst
Part C: a polymeric isocyanate, e.g. polymeric methyl di-isocyanate (MDI)

Part A is normally used at between 1 and 2% of the sand weight, with Part B added at between 2 and 10% of the weight of Part A, and Part C added at between 18 and 20% of the weight of Part A. Two distinct stages are involved in the curing action. Firstly, the polymeric isocyanate (Part C) reacts with the alkyd resin (Part A) at a rate controlled by the amine catalyst (Part B). This reaction produces a urethane bond which provides sufficient strength in the mould for the pattern to be stripped. The second stage of cure involves the oxidation polymerisation of the alkyd resin by atmospheric oxygen and is accelerated by the presence of the metallic elements associated with the Part B component. This stage is relatively slow at room temperature and is a function of the section thickness of the sand. Heating to between 150 and 205°C (300 to 400°F) will accelerate the second stage reaction.

***Phenolic Urethane System.*** These binders were introduced during the early 1970's and are suitable for the production of ferrous and non-ferrous alloys. Formulations can be tailored to suit jobbing or high-production operations. The three components of this binder system are:

Part I: a phenol formaldehyde resin dissolved in a blend of solvents
Part II: a polymeric isocyanate (e.g. MDI) dissolved in a blend of solvents
Part III: a liquid amine catalyst

The phenol formaldehyde resin provides hydroxyl groups which combine with the isocyanate to produce a urethane bond at a rate dependent upon the strength and amount of amine catalyst. Unlike the alkyd urethane system there is no second stage reaction and a deep through cure is obtained. Binder levels of 0.7 to 2.0% of the sand weight are recommended, with a ratio of 55:45 or 60:40 of Part I to Part II [3]. The amount of catalyst required would be between 0.4 and 0.8% of the weight of the Part I addition. The offsetting of the ratio of Part I to Part II binders, in favour of the Part I binder, minimises the amount of nitrogen in the system. Improvements in casting surface finish, particularly for steels, can be obtained by small additions of iron oxide to the sand [3].

***Polyol Urethane System.*** This is a derivative of the previous system in which the Part I binder is a polyol which provides good thermal breakdown after casting. Part II is similar to that used in the previous system and the Part III catalyst may be dispensed with unless fast stripping times are required. This system was introduced in the late 1970's to extend the phenolic urethane system to the production of light alloys.

The commercial products are sold under the names of Pepset and Novathane [2, 4]. In recent years these binders have been modified to reduce their smell on mixing, smoke emission and lustrous carbon generation.

### Cold Box Processes

***Phenolic Urethane Cold Box System.*** This is the original organic binder, cold box system. It was introduced during the late 1960's and is known colloquially as the Ashland or Isocure process. It is similar to the phenolic urethane cold set system. The essential difference is that the sand is mixed with the Part I and II binders only before consolidation. The part III amine catalyst is introduced as a vapour through the consolidated sand to produce a rapid cure. Two amine catalysts are in common use: dimethyl-ethylamine (DMEA) and triethylamine (TEA). The former is more easily vaporised and is more reactive but it has the more pungent odour.

The process has found widespread application in both long-series and automatic production of cores and in the manufacture of smaller quantities of jobbing-type cores of modest size [5]. It has also been used in the production of boxless moulds. The major advantage of the process is its ability to produce dimensionally accurate moulds and cores with fast cycle times. Methods have been developed to reduce the amount of amine, the most expensive part of the system, used in the process [2]. This has involved more accurate metering or the use of a warm-box approach.

***The $SO_2$ Process.*** This process was introduced during the mid 1970's and is often referred to under various trade names such as So-Fast, Hardox and Insta-Draw [5]. The system is based on the use of a furan resin, with furfuryl alcohol/formaldehyde giving the best results. The sand is mixed with a peroxide, for example methyl ethyl ketone peroxide (MEKP) added at between 20 and 50% of the resin weight, an adhesion promoter and silane. Following compaction, sulphur dioxide gas is passed through the mixture and is oxidised by the peroxide to produce sulphur trioxide. The sulphur trioxide dissolves in the water associated with the resin to produce sulphuric acid which polymerises the resin.

The mixed sand has a long bench life, is free flowing, sets rapidly, and the moulds/cores have a good storage life [5]. However, problems associated with the process include: resin build-up on tooling, corrosion of ferrous equipment, sulphur pick-up in SG iron, and the danger of sulphur dioxide escaping into the atmosphere.

***The Epoxy-$SO_2$ Process.*** Introduced commercially in 1983 and known also under the trade name Rutapox, this process was developed to remedy the problem of resin build-up associated with the furan process [2, 6]. A two-part binder system is used, with Part A comprising a combination of a modified epoxy resin with an aromatic peroxide and Part B consisting of an epoxy resin with acrylic modifiers to improve reactivity and cure speed [6]. Polymerisation of the resin is achieved by the passage of $SO_2$ through the compacted sand. Problems of sand wash associated with this system have been overcome by modifying the epoxy resin with a phenol formaldehyde addition [2].

***The Free-Radical Cure Process.*** Introduced during 1982 and known as the FRC process, this process uses a three-component binder addition. The components are: i) a vinyl unsaturated urethane oligomer that contains reactive carbon-carbon bonds; ii) an organic peroxide; and iii) a vinyl silane adhesion promoter, to increase strength and improve storage life [6, 7]. Curing is effected by passing $SO_2$ gas, diluted to between 1 and 10% in a nitrogen carrier, through the compacted sand. The $SO_2$ gas breaks down the peroxide to produce free radicals which initiate the addition polymerisation reaction to cure the resin. The absence of nitrogen and water in this system is advantageous but the material can be susceptible to sand erosion, with ferrous alloys necessitating the application of a refractory mould coating [2].

***The Ester Cured Phenolic Process.*** This process was introduced in 1983 under the trade name "Betaset". The chemistry of the system is similar to that of the cold box version. The sand is mixed with a water-soluble alkaline phenolic resole with a typical addition level of 1.5% of sand weight. After compaction methyl formate vapour, carried by air, is passed through the sand, whereupon an addition polymerisation reaction occurs to cure the resin. Although the cured strength may not be as high as that produced by the urethane cold box system, the Betaset process finds widespread use in both moulding and coremaking applications. The system is particularly suitable for making boxless moulds as a replacement for green sand [2]. An example of a core produced by this process is shown in Fig. 3.2.

The original Ecolotec binder system used a similar approach to binder chemistry and curing [5]. However, this system has been developed to enable it to be cured using carbon dioxide gas [2]. The phenolic resole system is complexed with an oxyanion such as a borate. A conventional phenol formaldehyde resole is mixed with an alkali, such as potassium hydroxide, and borax is added to provide a stable solution. When carbon dioxide is passed through the compacted sand it dissolves in the water in the system to produce carbonic acid. The excess alkali in the system is neutralised by the carbonic acid and a complex then forms between the resole and the borax which gels the resin.

Fig. 3.2 Photograph showing a core produced by the Borden Betaset process

***The $CO_2$-Polyacrylic Process.*** In the Polidox process between 3 and 3.5% of a sodium polyacrylate solution, together with 1 to 1.5% of calcium hydroxide powder, are mixed with dry sand [2]. The passage of $CO_2$ gas through the compacted sand effects curing by promoting a reaction between the calcium hydroxide and sodium polyacrylate.

***The Redset Process.*** This three-component system consists of: a) a liquid resorcinol formaldehyde polymer; b) an acid catalyst (sulphonic acid); and c) methylal (a volatile acetal). The acid and polymer are mixed with the sand, which is blown into a warm box at 40°C and gassed with the methylal vapour [2].

Although this section is concerned primarily with the cold set and cold box systems, for the sake of completeness details of the hot box and warm box processes are included.

## The Hot Box Processes

The two processes which can be defined as hot box processes are the shell core-making and hot box core-making processes. The former is considered in the section dealing with the shell moulding process. In this section consideration is restricted to the hot box process.

***The Hot Box Process.*** The resins used in this process are based on furfuryl alcohol and/or phenol and invariably contain urea and formaldehyde. An acid catalyst is incorporated in the sand mixture with the resin and the resultant wet mixture is blown into a heated core box. Although the temperature will be determined by the resin formulation, it is typically within the range of 220 to 245°C (425 to 475°F). The application of heat induces a rapid cure, enabling cores to be removed within 10 to 30 seconds. Curing continues due to the action of residual heat, together with that generated exothermically from the polymerisation reaction. However, post-baking may be practised to ensure a complete cure [3]. Hot box cores have been used extensively by the automotive industry for the production of intricate cores with good tensile strength properties.

***The Warm Box Process.*** The term warm box was originally applied to a process first introduced in 1978 as an alternative to the hot box processes [8]. The resins used are cold blended furfuryl alcohol monomers which are phenol-free and low in nitrogen and water. Improved dimensional stability and strength are achieved with binder additions typically 20% below those for the hot box process. The catalysts are copper salts based primarily on aromatic sulphonic acids added in amounts between 20 to 35% of the binder weight [3]. They promote through curing at a temperature which is typically within the range 150 to 220°C (302 to 428°F).

## Mould Preparation for Casting

Cured moulds are coated with a refractory wash, if required, and the cores located prior to closing. Across-the-joint location for boxed moulds is achieved by conventional pin and bush means, but special provision must be made for boxless moulds.

## Gating and Feeding

The normal principles of gating and feeding apply to the production of castings by the cold set processes. As with the $CO_2$/Silicate process, the rigid mould may enable feeding requirements to be reduced for cast irons of appropriate composition.

## Casting Finishing

Finishing procedures are similar to those described for green sand moulded castings.

## Reclamation

Increasing costs for new sand and used sand disposal have encouraged foundries to adopt reclamation systems. These are predominantly based upon an attrition process which reduces the majority of the residual binder to a dust which can be separated from the sand. However, it is impossible to eliminate the binder entirely and consequently there can be a build up of elements which are environmentally undesirable or adversely affect the quality of the castings. In this latter respect the build up of sulphur, nitrogen and phosphorus is undesirable [2].

## Health and Safety Aspects

Considerable care must be taken when using organic binder systems in the handling of ingredients, sand mixing, moulding, and at the pouring and knock-out stages. Well ventilated working areas are an essential requirement and the use of protective clothing advisable. To ensure a safe working environment the most hazardous materials are processed automatically in fully enclosed machines supplied complete with their own environmental cleaning facility. Environmental standards also apply to the effluent from scrubbers and the disposal of waste sands.

**Process Capabilities**

***Alloy Range.*** The processes, with appropriate selection, are suitable for a wide range of alloys including aluminium- and copper-base, cast irons and certain steels.

***Casting Size and Weight Range.*** The processes are suitable for the series production of castings at the top end of the weight range suitable for green sand castings, the jobbing production of medium size castings and the production of large castings. In addition, processes may be selected for core production.

***Production Volumes and Lead Times.*** As indicated, a process can be selected to suit the volume of castings required. Lead times are similar to the $CO_2$/Silicate process, as are production rates.

***Thin Section Capability.*** As sand casting processes, a section of 5 mm (0.20 in) would be an acceptable minimum value.

***Detail Replication.*** As sand casting processes, expectations should not be high in this respect.

***Surface Finish.*** A surface roughness value in the range of 6.3 to 25 μm (250 to 1000 μin) can be expected.

***Cast Holes.*** Hole sizes should be restricted to $\phi$ 6 mm (0.25 in) and above.

***Dimensional Accuracy.*** Whilst the accuracy of castings is improved by the use of these processes, machining will still be required.

***Typical Castings.*** A wide range of castings for various engineering applications is produced by these processes and includes machine tool castings and castings for a whole range of mechanical plant and equipment.

An example of the successful exploitation of the cold box process for the production of precision sand castings is provided by the Cosworth process. However, although the cold box process is central to the Cosworth process, it is only part of a total system engineered to optimise casting quality. It is, therefore, worth considering this process in detail.

## THE COSWORTH PROCESS

### Introduction

The Cosworth process is a high-precision, low-pressure sand casting process which was developed for the production of premium-quality aluminium alloy castings for the automotive industry. Unable to obtain castings that could consistently meet their exacting standards of quality, Cosworth established a research and development programme to study the fundamental problems in moulding, melting, casting and heat treatment of aluminium alloy castings [9].

Commercial-quality aluminium alloy castings produced by sand or gravity die castings contain porosity and inclusions which impair their metallurgical integrity and reduce the attainable level of mechanical properties. Hydrogen is one source of porosity and shrinkage another. Inclusions arise primarily from the presence of alumina, which forms readily on the surface of aluminium and is a particular problem during the air-melting of aluminium alloys. These tenacious films of oxide are readily entrained in the melt during metal transfer, pouring and movement through the gating system. The presence of this alumina as particles or films within the casting structure produces discontinuities which reduce mechanical properties. Furthermore, the oxide particles provide interfaces upon which gas or shrinkage porosity can nucleate.

Inaccuracy is an inherent problem of castings produced in silica sand moulds and is a direct consequence of the polymorphic change of quartz from its $\alpha$ to $\beta$ form at 573°C (1063°F). The volumetric expansion associated with this phase change causes moulds and cores to expand in a manner which is unpredictable. This unpredictability is reflected in the dimensional variability of the castings produced. Even if a metal die is substituted for the sand mould, as would be the case in gravity die casting, the use of sand cores would still create a degree of dimensional inaccuracy because of the differential expansion between the die and core.

Conventional castings may also exhibit other troublesome, if less fundamental, features which include [9]: blowholes from chills, cores and adhesives; inaccurately located cores and mould halves; metallurgical inadequacies (particularly poor hardness or strength); and the considerable fettling required to remove large feeder heads and extensive gates.

To overcome these problems Cosworth developed a complete process, engineered from first principles, to produce castings for the Cosworth engine in 1978. The process was extended to the largely-automated series production of commercial castings with the opening of a new foundry in 1984 [10]. In 1986 Cosworth Research and Development Ltd. received the Queen's Award for Technological Achievement for the casting process. The most recent accolade has been provided by the Ford Motor Company which has selected the process for its Windsor Ontario plant, due to open in mid-1993 [11]. The advantages claimed for the Cosworth process include [14]:

- considerable weight saving, allowing design of lighter and more robust components.
- exceptional high strength and ductility.
- minimum machining.
- dimensional accuracy which can exploit modern manufacturing processes.
- components which have total commitment to value engineering.
- castings which are pressure tight due to the absence of porosity and inclusions.
- improved metallurgical consistency.
- flexibility to meet changing markets.
- comparatively inexpensive tooling.
- suitable for medium to high volume production.
- high metal yields obtainable.

## Process Outline

***Mould Production.*** The starting point for the production of an accurate casting is accurate pattern and core box equipment. Resin pattern equipment is suitable and permits any parts which are susceptible to wear to be fitted as replaceable inserts which can be recast from a master [9]. To overcome the problem of unpredictable expansion associated with silica sand, zircon sand was chosen. To enhance the surface finish of the castings, careful choice was exercised over sand grading and quality and the chemical binder system [12]. The choice of a cold set process eliminated the problems associated with clay-bonded sands. It also eliminated the inaccuracies associated with hot processes, as there is no distortion caused by thermal expansion of either core or core box [9]. The gas-hardening $SO_2$ process was chosen [12]. The total binder content required is low and this reduces the danger of blowholes caused by the decomposition of the binder when subjected to heat from the metal. The moulds and cores are produced by automatic core shooters and the cores are manually placed in the drag moulds. Mould closure is automatic, with location effected by a pin and bush system accurate to within ± 0.05 mm (± 0.002 in) [9]. The closed mould is transferred to a pick and place unit which positions it on the casting station above a sealed holding furnace [13].

***Metal Preparation and Transfer.*** The metal is melted and held in an electric resistance heated melter-holding furnace which is unpressurised and has a sufficient capacity to even out small fluctuations in analysis. The dwell period in the furnace is sufficient to enable non-metallic particles to separate by floating or sinking and metal for casting is drawn from the middle zone of the metal from which the particles have separated. Controlled degassing is applied at the ingot charging station and a protective atmosphere is used to minimise the risk of gas absorption and oxide formation during holding [14]. No fluxing or chemical treatment is applied to the metal and this includes no modification or grain refinement.

The metal is transferred by a pump from the holding furnace to the mould and enters from below in a quiescent manner. A magnetic pump, which has no moving parts and works on the principle of a linear motor, can be used to fill the mould under predetermined, controlled conditions of pressure and rate [9]. Alternatively, a pressure pump which uses inert gas to discharge the metal from the holding furnace to the mould may be used [15].

No feeder heads are used, although there is some pressure-assisted feeding through the gates. The maintenance of pump pressure until solidification is complete presents a production constraint, although cycle times are equivalent to those in gravity die casting and low pressure die casting of equivalent components [14]. However, this constraint has been removed by a process modification in which the mould is inverted through 180° after filling [15]. The inversion is completed with the pressure still applied. However, after inversion the mould can be separated from the pump and transferred to a cooling station whilst the next mould is connected to the pump. Although production rate is increased it is at the expense of a reduction in yield, as an auxiliary or feed supply of metal is required.

***Metallurgical Considerations.*** As indicated, aluminium-silicon alloys are not modified nor are they grain refined. Evidence suggested that modification was not advantageous as the change in structure appeared to be accompanied by an increase in dissolved gas and heterogeneous nuclei which led to an increase in microporosity [17]. Dendrite arm spacing (DAS) is improved by the use of a zircon sand, which encourages a higher rate of heat transfer compared with silica sand. Values for DAS of 30 μm (0.0012 in) are obtained by the process, which compares with 60 to 120 μm (0.0024 to 0.0048 in) for conventional sand or investment casting and 20 to 50 μm (0.0008 to 0.0020 in) for gravity die castings [17]. The improved metallurgical integrity of castings produced by the Cosworth process is claimed to provide improvements in tensile strength in the order of 25% above the values obtained by conventional sand casting [9].

### Mould Preparation for Casting

Preparation of the moulds for casting follows the procedure common for boxless moulds produced by cold set processes. Coating of the mould with a refractory wash is unnecessary.

### Gating and Feeding

The procedures for gating and feeding castings are quite different from conventional sand casting practice. Metal enters the base through gates which connect the cavity to the metal distribution manifold. These gates, in association with the pump control system, control the rate of mould filling and provide pressure assisted feeding. The slowest casting rates compatible with reliable filling of the cavity are used, for example around 1 kg/s (2.2 lb/s) for complex cylinder heads [17].

### Casting Finishing

The use of a fine grained zircon sand, and the absence of conventional gating and feeding systems, reduces the fettling requirements, which otherwise are conventional.

### Reclamation

Cosworth has developed a special reclamation process suited to the material and binder, it is based on thermal reclamation.

## Health and Safety Aspects

The general points outlined in the section on cold set processes apply to the Cosworth process.

## Process Capabilities

***Alloy Range.*** The process has been developed for the production of aluminium alloys but there is no reason, in principle, why it should not be applied to other non-ferrous alloys. Indeed, it could be used for ferrous alloys if a pressure pump rather than a magnetic pump was employed.

***Casting Size and Weight Range.*** The weight range quoted for aluminium alloy castings is 0.2 to 55 kg (0.44 to 121 lb). The maximum mould size used in the Cosworth Castings Ltd. plant was 915 x 610 mm (36 x 24 in).

***Production Volumes and Lead Times.*** Although the patterns are required to be accurate, the lead time for pattern production need be no greater than is usual for resin pattern production. The process has now been refined to the extent that it can, from a single set of tooling, produce castings at a rate considerably in excess of that possible with either the gravity or low-pressure die casting processes.

***Thin Section Capability.*** Although the quiescent filling of inclusion-free metal and pressure-assistance are advantageous, the high heat capacity of the moulds is a disadvantage. Sections of 2.5 mm (0.10 in) have been produced in castings with an overall dimension of 200 mm (8 in). However, a value of 4 mm (0.16 in) would be more realistic in production castings.

***Cast Holes.*** No data available, but hole diameters of at least $\phi$ 6 mm (0.25 in) are possible.

***Surface Finish.*** The surface finish of castings produced by the process is better than conventional sand casting but cannot match investment casting. A typical range for surface roughness would be 3.2 to 6.3 μm (125-250 μin).

***Dimensional Accuracy.*** Attainable accuracy, in relation to size of dimension, is shown in Table 3.1.

Table 3.1 *Attainable Dimensional Accuracy in Castings Produced by the Cosworth Process*

| Dimension Size | | Tolerance | |
|---|---|---|---|
| mm | in | mm | in |
| up to 100 | 4 | ± 0.15 | ± 0.006 |
| up to 300 | 12 | ± 0.20 | ± 0.008 |
| up to 800 | 32 | ± 0.25 | ± 0.010 |

An additional allowance of ± 0.10 mm (± 0.004 in) should be added for each across-the-joint or core joint dimension.

***Typical Castings.*** Castings produced by the Cosworth process for the aerospace and defence industries include: gas turbine front end components, fuel system pumps and controls, flight refuelling manifolds, weapon mountings, lightweight undercarriage components, manifolds and ductings, and transmission housings. Automotive castings include: high performance cylinder heads for racing engines, 16 valve cylinder heads for passenger cars, air cooled cylinder heads, marine cylinder heads, engine blocks, sumps and transmission cases [14].

## THE SHELL MOULDING PROCESS

### Introduction

The shell moulding process is a precision sand casting process capable of producing castings with a superior surface finish and better dimensional accuracy than conventional sand castings. These qualities of precision can be obtained in a wider range of alloys and with greater flexibility in design than die casting and at a lower cost than investment casting. The process was developed and patented by Croning in Germany during World War II and is sometimes referred to as the Croning shell process.

The fundamental feature of the process is the use of a fine-grained, high purity sand which contributes the attributes of a smooth surface and dimensional accuracy to moulds, cores and castings alike. In conventional sand moulding the use of such a fine sand is precluded because it would dramatically reduce mould permeability. This has the effect of retarding the escape of air and mould gases, causing short-run castings or castings containing gas defects. However, the distinguishing feature of the shell moulding process is that the mould is literally a shell, being in the region of only 10 mm (0.4 in) thick. It was the ability to produce such a thin shell mould which made the process a revolutionary development in metal founding. The coincident development of plastics, like Bakelite, which were based on thermosetting resins such as phenol formaldehyde, provided the basis for shell moulding. In shell moulding the fine sand is coated with a thermosetting resin which provides the relatively high strength required to enable a thin section, or shell, mould to be produced.

The requirement that the mould should accurately replicate the pattern detail and dimensions if a precision casting is to be produced, is also met by the shell moulding process. This is achieved because the resin bond is developed whilst the mould is in contact with a heated pattern plate. Furthermore, the mould is separated from the pattern without the need to enlarge the cavity, as is the case in green sand moulding. These features apply equally to the production of cores by the process. A further improvement in casting accuracy can be obtained if zircon sand is used instead of silica sand. This arises because the expansion of zircon sand, caused by the heat of the cast metal, is both lower and more predictable than that of silica sand. Foundry production of castings by the process is comparatively straightforward and the process lends itself readily to close control, with the advantage of consistency in the castings produced.

The advantages and disadvantages of the process are summarised below [18, 19].

Advantages

- lower capital plant costs, when compared with mechanised green sand moulding.
- capital outlay on sand preparation plant is not essential.
- good utilisation of space.
- low sand to metal ratio.
- mould coatings are unnecessary.
- lightweight moulds are produced which are readily handled and have good storage characteristics.
- skilled labour is not required.
- shells have excellent breakdown at the knock-out stage.
- lower cleaning and fettling costs.
- castings have a superior surface finish and dimensional accuracy, when compared with green sand moulded castings.

Disadvantages

- the raw materials are relatively expensive.
- the size and weight range of castings is limited.
- the process generates noxious fumes which must be effectively extracted.

**Mould Materials and Preparation**

***Sand.*** The ideal properties of a sand for use with the shell moulding process are: it should be fine with a narrow grading; the grading should be consistent; if siliceous, it should be of high purity to ensure high refractoriness; and the grains should be rounded and their surfaces free from impurities to minimise the resin addition required [20]. A sand with an AFS grain fineness number of around 100 would be typical of the sand grading specified for the process. In addition to being clean the sand must be dry, in order to ensure that resin coating and subsequent strength development is effective. Grain size and distribution affect permeability, surface finish, binder demand and mould strength. An objective of sand selection is the optimisation of these characteristics. Of the non-siliceous alternatives zircon is most widely used and, due to its high cost, would almost certainly be reclaimed. Olivine is difficult to coat and requires high resin additions [19].

***Resin.*** Phenol formaldehyde resins are used in shell moulding because, when combined with sand, they have high strength, resistance to heat and moisture, and good flowing and curing properties [21]. Although the basic resin type has not changed since the inception of the process, there have been many developments to modify the resins to produce faster investment and cure, higher strengths and resistance to peelback [10].

The resin used for the shell moulding process is described as a two stage or novolak resin. A novolak resin is manufactured by reacting an excess of phenol with formaldehyde under controlled acidic conditions to produce a phenol formaldehyde resin which has insufficient formaldehyde present to enable the resin to cure as a thermoset. In fact, in this condition, the novolak behaves as a thermoplastic in that it may be repeatedly melted without changing its properties. In order to cure the resin a curing agent must be included in the sand formulation.

***Curing Agent.*** Hexamethylene tetra-amine, commonly referred to as hexamine or hexa and produced by reacting formaldehyde with ammonia, can be used as the curing agent for shell sands. It breaks down rapidly at the shell curing temperature to release formaldehyde, which is then responsible for linking the novolak chains together to provide a rigid cross-linked, or thermoset, structure [20]. Unfortunately, the ammonia liberated from the hexa can react with, and become fixed in, the novolak structure. The associated nitrogen can produce nitrogen pin-hole defects in iron or steel castings. However, resin formulations have been developed to be nitrogen-free or low in nitrogen to overcome this problem [22].

***Additives.*** Lubricants are added to the sand-resin mixture for two reasons. Firstly, they improve the flowability of the sand, which improves packing density and consequently shell strength. Secondly the lubricant aids the separation of the cured shell from the pattern equipment. The most commonly used lubricant is calcium stearate, because of its efficiency and superior temperature resistance [20]. Other additions are made to reduce the tendency of shells to crack during pouring and to counter casting defects. Vinsol, derived from wood resin, is used to reduce shell thermal cracking tendency, which it does by plasticising the rigid phenolic structure [20, 21]. Iron oxide is an additive used to counter the orange-peel defect associated with certain grades of steel. Its addition with vinsol may also improve crack resistance [21].

***Sand Preparation.*** Pre-coated sand, complete with appropriate additives, may be purchased directly from the suppliers, obviating the need for capital plant in the foundry. However, users of large quantities of sand, and especially reclaimers of sand, prefer to mix their own. Templar [19] describes the four main methods of pre-coating sands with resins, which are: solvent method, plasticiser method, warm process and hot process. However, the superior productivity associated with the latter makes it the preferred procedure [20]. In the hot process sand, within the temperature range of 115 to 150°C (240 to 300°F), is charged into a batch mixer which may be a pug mill/paddle mixer or operate on the muller wheel principle [20]. Solid flaked or pelleted coating resin is added and, depending on the resin, is allowed to mix for between 45 and 75 seconds. Calcium stearate, hexamine and cooling water are then added, with mixing continuing for a further 60 to 80 seconds to allow cooling by water evaporation to take place. Finally the mixture is discharged into a disintegrator, screened and cooled in a fluidising bed [19]. The amount of resin used in the process is between 2 and 5% based on the weight of the sand. The actual amount will depend upon the sand grading and quality and the shell properties required. If used, the hexa addition would be about 15% by weight of the resin addition. Vinsol additions of around 10% by weight of the resin addition and stearate additions of between 2 and 5% by weight of the resin addition are recommended [21]. The pre-coated sand is dry and free-flowing and does not readily absorb moisture from the atmosphere. It can be conveyed pneumatically and stored in hoppers.

## Tooling Requirements

In order to obtain an accurate casting a precision pattern must be used. As the pattern must be capable of withstanding repeated heating to temperatures around 300°C (570°F), only metal patterns can be used. Whilst several materials have been tried, cast iron has been found to be the most suitable. The material is easily machined and, providing that it is stress relieved before final machining, distortion is unlikely to be a problem. The pattern equipment must make provision for the gating and feeding system, core location, mould part location, and for the ejector pins used to separate the cured shell from the pattern. To avoid distortion the pattern plate should be at least 25 mm (1 in) thick and although a minimum of taper is required on vertical surfaces, at least 0.5° of taper is advisable.

## Mould and Core Production

The pattern plate assembly is heated, predominantly by gas, to a temperature between 200 and 240°C (390 to 465°F), when it is invested with the shell moulding sand. The roll-over, dump-box technique is the original method of investment, but the fluidised sand method and controlled contour method are also used.

***Dump-Box Method.*** In this method the heated pattern plate is clamped to the dump-box which contains the sand, see Fig. 3.3. The whole assembly is rotated through 180° so that the sand falls into contact with the heated pattern plate. The investment time will depend upon the temperature, the resin and the required thickness of shell, and typically would be between 10 and 30 seconds. During this period the resin softens to provide a viscous liquid bond between the sand grains. At the end of the investment period the assembly is returned to its original position and the pattern plate is released. Any sand which was not heated to a temperature sufficient to soften the resin simply falls back into the dump-box for use in succeeding cycles. The thin layer of sand coating the pattern must now be heated to a temperature in the region of 260 to 300°C (500 to 570°F) to decompose the hexamine, if present, to effect cross-linking or curing of the binder. Curing will take place within 1 to 3 minutes, depending upon the temperature, resin and shell thickness. After curing the shell is mechanically separated from the pattern plate by ejector pins which operate through the plate.

The dump-box method described has two limitations, which arise because the sand rolls across the pattern as the dump-box rotates. This results in reduced compaction on the shadow side of the pattern, reducing shell strength and increasing the tendency for peelback to occur when the pattern plate is separated from the dump-box. Techniques to overcome this problem include: rotation at high speed, which effectively drops the sand vertically onto the pattern; the use of a louvre, which retains the sand until rotation is complete and is then removed to allow the sand to fall vertically onto the pattern plate; and the use of a fluidised bed, in which the sand is fluidised in a vertically upwards direction to coat the pattern plate.

***Controlled Contour Method.*** In this method the moulds are produced by blowing the sand between the heated pattern plate and an associated heated, profiled back-plate. This back-plate may incorporate a ribbed configuration to provide improved mould rigidity and casting support. Despite higher pattern costs, this method is favoured because it offers close control over the shell thickness which, in turn, leads to better casting consistency and quality.

***Shell Thickness.*** As indicated, the shell thickness produced will be a function of pattern temperature and investment time. However, the thickness of shell required will be a function of its application which, in turn, will depend upon: the metal to be cast, its density and temperature; the orientation of the mould during pouring; and whether or not support is provided to the mould during pouring. Apart from these practical aspects and the implications for casting quality, there is the aspect of cost to consider. Thicker moulds require more material, more energy and take longer to produce, thus increasing the cost of manufacture. An investigation to produce a process model for the shell moulding process showed that shell thickness can be predicted as a function of different operating parameters and binder formulations [23]. This may lead to process optimisation in terms of cycle-time and energy reductions.

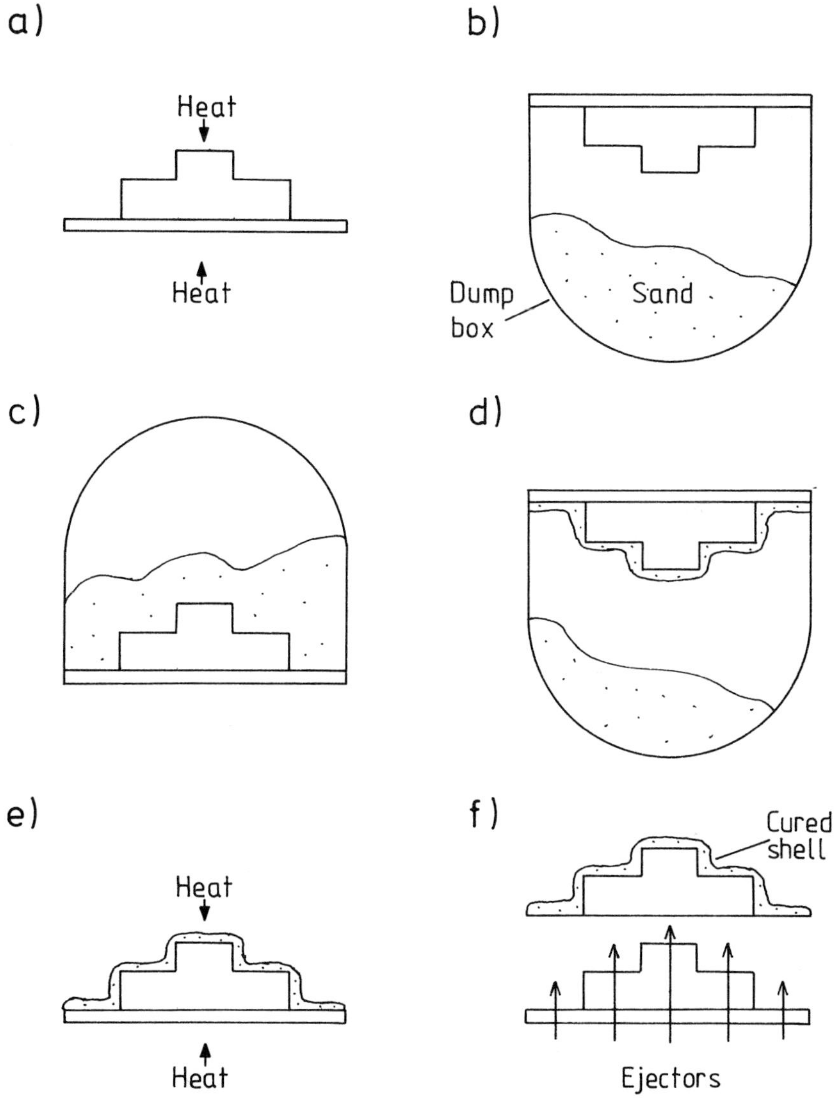


Fig. 3.3 Schematic representation of the stages in the manufacture of a shell mould by the dump-box method: a) the plate and pattern are heated; b) the plate assembly is clamped to the dump box; c) the whole assembly is rotated through 180° so that the sand contacts the heated plate; d) the assembly is returned to the original position; e) the sand layer is heated to effect curing; f) the cured shell is ejected from the pattern plate.

***Curing.*** Control of curing is important, as undercured moulds will be weak and generate excessive gas during casting. Overcuring may result in the moulds breaking down prematurely and eroding under the action of the metal. In a study of novolak resin chemistry it was established that, when using hexamine as the curing agent, cross-linking commenced at 150°C (300°F) and that hexamine undergoes extensive decomposition at 260°C (500°F) [24]. The two variables over which close control must be exercised during curing are time and temperature.

***Mould Stripping.*** In order to ensure that the moulds are separated from the pattern without damage or distortion, to which they are particularly prone when hot, careful attention must be paid to the ejector design. The most common method of ejection uses ejector pins which push upwards through the pattern plate to lift the mould clear of the pattern. Sufficient pins must be used to ensure a good strip. A pin should be placed close to each corner of the mould to prevent warping. Others should be spaced more or less evenly over the plate surface [19].

***Shell Core Making.*** The attributes of shell moulding can be conferred on cores by the shell core making process. Indeed, the advantages of shell cores are not restricted to their use in shell moulds, as they are widely used in green sand moulding and gravity die casting. To produce a shell core, the coated sand is blown into a heated metal core box and allowed to invest for a short period of time. Uncured sand is removed, usually by inverting the core box so that the sand simply falls out, and the invested sand is cured. The cured, and generally hollow, core is extracted and the cycle repeated.

**Mould Preparation for Casting**

The moulds are assembled, incorporating cores if required, using adhesives applied to the mould joint surface. The amount of adhesive must be controlled to avoid excessive growth in dimensions across the mould joint line. Adhesives are available to suit the assembly of hot mould halves, cold mould halves, or hot to cold mould halves. Pressure is usually applied at the closing stage to improve adhesion and minimise joint line growth. In vacuum mould closing, the mould is completely enclosed between two rubber diaphragms and exhausting air from between them exerts uniform pressure over the mould surface, irrespective of its contour. A depth of contour up to 155 mm (6 in) can be accommodated, above this pin-type, press closers are preferred [19].

Shells may be poured without support but, because they are thin and their strength deteriorates due to the action of the metal's heat, it is recommended that they be supported during and after pouring. This reduces the tendency for shells to distort or crack, either of which would impair the quality of the castings. Green sand, cold set sands, metal shot and granular refractory material have all been proposed and used for supporting shell moulds. However, the latter is the most practicable. Jupp [18] has described the procedure for processing supported moulds and highlighted the advantages as follows:

- sand costs may be reduced as a lower resin content is acceptable and less sand is used.
- shell production rate is increased as thinner shells can be used.
- mould dilation is eliminated.
- mould burst-out is eliminated.
- less fume and flame are evolved during pouring.

***The Vacustract Process.*** Researchers at SCRATA [25] developed an improved shell support system with the added advantage of fume and flame suppression. In this process the shells are immersed in a bed of fluidised granular material. During pouring, and for a period thereafter, extraction is applied in such a way that air passes from the atmosphere through to the base of the bed, where it is removed, see Fig. 3.4. This approach has three principal advantages:

(i) The pressure differential created increases the resistance of the bed to shear forces, thereby enhancing the support that the bed provides, which considerably reduces the risk of mould failure.

(ii) The extraction draws emissions from the mould through the bed, rather than allowing them to escape to atmosphere, thus controlling the immediate environment and suppressing the formation of flame.

(iii) The continuous flow of air reduces the amount of heat transferred to the bed material from the castings, thus reducing the need for subsequent cooling.

This approach is applicable to moulds produced by other processes and is fundamental to the successful application of the Replicast CS process.

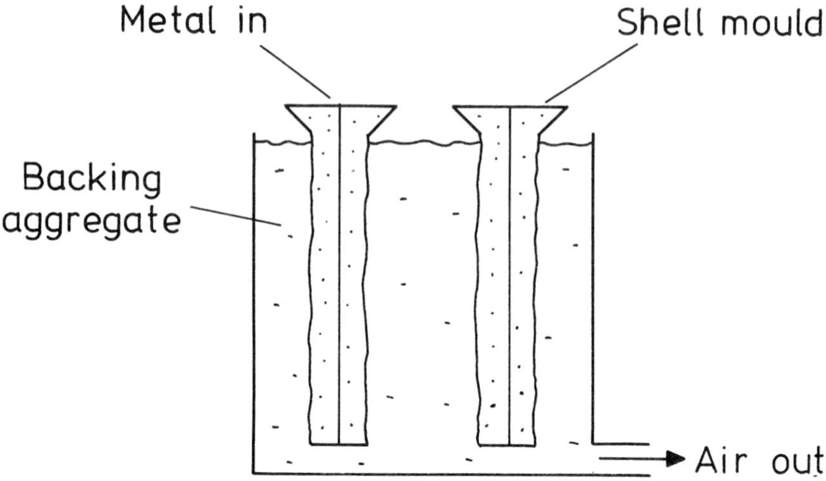


Fig. 3.4 Schematic diagram of the Vacustract process

***The CLAS Process.*** Shell moulds are normally poured under gravity and, as they are permeable, are subject to the problems of turbulent filling, aspiration, and the associated problems of gas absorption and inclusions. The CLAS process applies the concept of counter gravity casting to the production of a shell mould [26]. In this process the castings are gated through the base of the mould directly into the cavity or a blind feeder. Mould filling is achieved by attaching the top of the mould to a vacuum chamber and submerging the mould base into a bath of molten metal, see Fig. 3.5. The metal is drawn into the mould by the vacuum suction, which also removes the fume generated by the mould. Improved casting yield and metallurgical integrity are claimed and the castings are produced at a cost lower than that for conventional shell moulded castings. No evidence has been found of melt contamination from the shells and the problem of gas defects in certain steels has been overcome by a patented technique which incorporates gas collection voids in the mould [27].

**Gating and Feeding**

In general, the normal principles of gating and feeding apply to the production of castings by the shell mould process. The fine sand grains and exothermic effects of resin burn-out improve fluid flow and the filling of thin sections. However, the chilling nature of zircon sand may counter these advantages.

Vertical or horizontal gating systems may be employed, according to the orientation of the mould for casting. Feeders may not be required for cast irons having an appropriate composition because the moulds can be considered to be rigid.

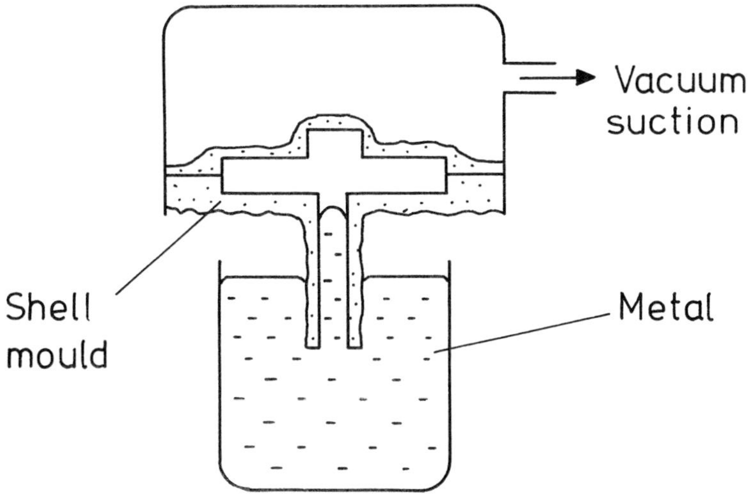


Fig. 3.5 Schematic diagram of the CLAS process

**Casting Finishing**

Finishing is by conventional means, although the requirement for shot blasting is reduced because sand adherence is eliminated by the presence of lustrous carbon generated by decomposition of the resin.

**Reclamation**

Shell moulding sand can be reclaimed by a thermal process in which the sand is heated to burn out the organic materials. After cooling and screening the sand can be recoated with resin.

**Health and Safety Aspects**

At the mould making stage ammonia and formaldehyde are released from the decomposition of hexamine and at the pouring stage there is, in addition, the emission of phenol and carbon monoxide. These obnoxious and harmful chemicals must be effectively extracted from the working environment and the extracted air treated before release to atmosphere.

## Process Capabilities

***Alloy Range.*** The process is suitable for a wide range of alloys which include: aluminium- and copper-base, all grades of cast iron, SG iron, low and medium carbon steels, high alloy steels, and manganese steels.

***Casting Size and Weight Range.*** Castings weighing from a few grammes (ounces) to hundreds of kilogrammes (pounds) can be produced by the process, but most production probably lies within the range 0.5 to 10 kg (1 to 22 lb).

***Production Volumes and Lead Times.*** Relatively expensive, precision metal tooling is required for the process, which extends the lead time for production. However, once the tooling is established in the foundry, high production rates can be achieved. The process is better suited to series production but small runs are acceptable, providing that the increased cost of production is compensated by savings in machining.

***Thin Section Capability.*** With fluid metals, such as grey cast iron, sections of 1.9 to 2.3 mm (0.075 to 0.090 in) can be routinely cast. It is claimed that sections of 1 mm (0.040 in) can be cast in iron and steel components produced by the CLAS process.

***Detail Replication.*** The use of a fine sand, a mould material which sets in contact with the pattern, and improved fluid flow characteristics ensure detail replication that is superior to conventional sand castings, although not as good as plaster or ceramic moulded castings.

***Surface Finish.*** As a general guide, the surface finish attainable by shell moulding is of the order of 3.2 μm (125 μin), which compares favourably with the average value for sand castings at 12.5 μm (500 μin).

***Cast Holes.*** Holes of 3 mm (0.125 in) can be cast by the process.

***Dimensional Accuracy.*** Attainable accuracy, in relation to size of dimension, is shown in Table 3.2. An allowance of ± 0.25 mm (± 0.010 in) should be added to dimensions measured across the mould joint line.

Table 3.2 *Attainable Dimensional Accuracy in Castings Produced by the Shell Moulding Process*

| Dimension Size mm | Dimension Size in | Tolerance mm | Tolerance in |
|---|---|---|---|
| up to 75 | up to 3.0 | ± 0.38 | ± 0.015 |
| 76 to 180 | 3.1 to 7.0 | ± 0.51 | ± 0.020 |
| 181 to 230 | 7.1 to 9.0 | ± 0.64 | ± 0.025 |
| 231 to 305 | 9.1 to 12.0 | ± 0.76 | ± 0.030 |
| 306 to 406 | 12.1 to 16.0 | ± 0.94 | ± 0.037 |
| 407 to 510 | 16.1 to 20.0 | ± 1.20 | ± 0.047 |

Fig. 3.6 Photograph showing a selection of castings produced by the shell moulding process

***Typical Castings.*** The range of applications is very wide for shell moulded castings. It includes castings for wear applications, valves, the machine tool industry, hydraulic equipment, automotive applications, cast gears, etc. A selection of castings produced by the shell moulding process is shown in Fig. 3.6.

## THE COLSHELL PROCESS

As its name suggests, this process is a combination of the cold set process and shell moulding. Contoured moulds are produced from a sand mixture which hardens in contact with the pattern without the application of heat. To achieve an acceptable rate of output, a fast reacting cold set system must be used. The process was developed by BCIRA in partnership with Spenklin Polygram Ltd. and the totally enclosed structure of the machine ensures its suitability for the Isocure and $SO_2$ systems.

In this method of mould production the sand mixture falls from a hopper into a shallow frame which has a flexible diaphragm base. The frame containing the sand is pressed vertically upwards against the pattern and pressure and vibration are used to consolidate the mould material. Whilst in this position the mould material is hardened by the passage of the reactive gas to produce a contoured shell mould. The mould would typically be 1.5 x thicker than a hot shell mould, but its production requires less binder and energy [28].

Wood and resin patterns are suitable for the process and an output of 90 mould halves/hour can be achieved in the production of moulds with a size of 610 x 405 mm (24 x 16 in) [29]. It is claimed that the process provides a sand saving of 65% over conventional cold set sand moulds. The moulds can be produced for casting in the vertical or horizontal plane. Figure 3.7 shows Colshell moulds being poured.

## THE ZEUS PROCESS

Although the Cosworth process radically improved the precision attainable in sand castings, the use of zircon as a mould material limited the minimum section of casting that could be attained. The Zeus process reverts to the use of silica sand in the production of thin section, precision sand castings. For such castings, produced in aluminium alloys, the heat input to the sand is insufficient to raise the sand's temperature to the $\alpha$ to $\beta$ quartz inversion temperature.

Consequently, for such thin section castings, precision is retained. Casting sections in the range of 2 to 2.5 mm (0.08 to 0.10 in) are claimed to be routinely possible for this casting process.

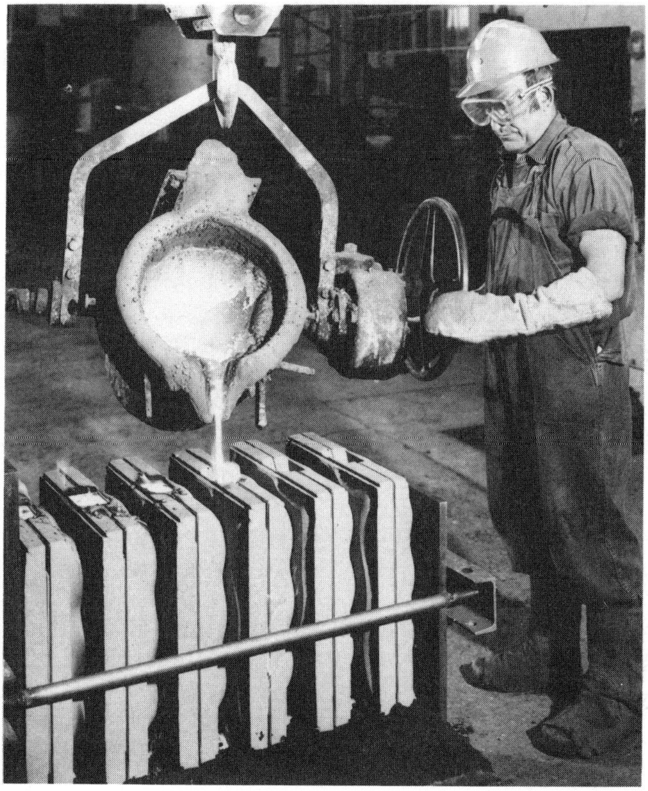

Fig. 3.7 Photograph showing the pouring of Colshell moulds

## References


1. Carey, P. R. and Schneider, J. T. (1986). Updating binder processes - part IV, *Foundry*, **114**, 5, 62-66.

2. Lemon, P. H. R. B. (1989). Modern sand binders, a review. *Foundry Trade Journal,* **163**, 3403, 922-924, 926.

3. Archibald, J. J. and Smith, R. L. (1988). Resin binder processes. In: *Metals Handbook, Vol. 15 Casting*, 214-221. Published by ASM.

4. Busby, A. D. (1989). Novathane - a process for the '90's. *Proceedings of the SCRATA Annual Conference*, Paper No. 1, 11 pp.

5. Bewick, T. (1985). Gas-hardened sand-bonding technology - a status report. *Foundry Trade Journal*, **158**, 3299, 61, 63-66.

6. Carey, P. R. and Archibald, J. J. (1986). Updating resin binder processes - part II. *Foundry*, **114**, 3, 53-54, 56-57.

7. *Chemically Bonded Cores and Molds.* Published by the American Foundrymen's Society, Des Plaines, Illinois, USA, 1987.

8. McKillip, W. J. (1987). The warm box process; an alternative core production process for castings. *The Foundryman*, **80**, 11, 456-460.

9. Wilkins, P. S. A. (1981). GKN + Cosworth - a new concept in aluminium casting. *Foundry Trade Journal*, **151**, 3223, 592-598, 606.

10. McCombe, C. (1986). Cosworth castings - right first time and every time. *Metals Industry News*, **3**, 3, 6-7.

11. McCombe, C. (1990). Enter Cosworth. *Foundry Trade Journal International*, **13**, 2, 3.

12. Campbell, J. (1981). Production of high-technology aluminium-alloy castings. *Engineering*, **221**, 3, 185-188.

13. Kellock, B. (1984). British technology tops German design. *Machinery and Production Engineering*, **142**, 3660, 20-22.

14. Lavington, M. H. (1986). The Cosworth process - a new concept in aluminium alloy casting production. *Metals and Materials*, **2**, 11, 713-719.

15. Campbell, J. (1983). A method and apparatus for low pressure casting of metals. *UK Patent 2103132A*.

16. Smith, R. A. (1987). Method of and apparatus for casting. *European Patent 0234877*.

17. Smith, R. A. and Wilkins, P. S. A. (1986). Low pressure sand casting: current experience with a new process. *Trans AFS*, **94**, 785-792.

18. Jupp, J. E. (1978). Economic use of the shell moulding process. *Foundry Trade Journal,* **144**, 3139, 1262, 4, 6.

19. Templar, B. J. (1978). Shell moulding - a review. *Ibid*, 1217-1250.

20. Terron, C. (1987). The shell process - 40 years of development. *Ibid,* **161**, 3342, 74, 76, 78, 79.

21. Shell molding. In: *Metals Handbook, Vol. 5B Melting and Casting,* 181-202. Published by ASM, 1972.

22. Lemon, P. H. R. B. and LeServe, F. L. (1967). Improvements in or relating to foundry moulding mixtures and a method of using same. *UK Patent 1077242.*

23. Aranguren, M. I., Borrajo, J. and Williams, R. J. J. (1984). Study of the formation of shell sand parts used in the shell molding process. *SAMPE J,* **20**, 3, 18-23.

24. Woo, C. C. and Hu, J. (1988). A study of phenolic novolak resin chemistry in relation to the hot method of coating shell sand. *Cast Metals,* **1**, 3, 171-175.

25. Bish, D. A., Sims, B. J. and Ashton, M. C. (1979). A system for casting shell moulds. *Proceedings of the 24th Annual SCRATA Conference,* Paper No. 4, 4 pp.

26. Chandley G. D. (1983). Automatic counter gravity casting of shell molds. *Modern Casting,* **73**, 10, 29-31.

27. Chandley, G. D. (1985). Gas permeable metal casting mold having gas collection voids. *UK Patent 2160133A.*

28. Fallows, J. and Harpin, P. J. (1979). The "Colshell" process. *Proceedings of the 24th Annual SCRATA Conference,* Paper No. 5, 10 pp.

29. Fallows, J. and McCormack, W. (1979). A new cold-shell moulding machine. *Foundry Trade Journal,* **146**, 3157, 458-468.

CHAPTER 4

# Precision Sand Casting Processes (Unbonded Sands)

## INTRODUCTION

The possibility of using a mould material which does not require a binder to provide mould rigidity during the metal pouring and casting solidification stages is very attractive for a number of reasons. If no binders are required then their cost is saved, as is the cost of mixing them with the sand and the cost of the capital equipment required for mixing. Reclamation will be simplified if there is no binder to be considered and maximum use can be made of the recycled material. Furthermore, if there are no binders or additions made to the moulding material, then many of those casting defects attributable to them will be eliminated.

This possibility can be achieved by using either one of the unbonded sand moulding processes, which are known respectively as the Evaporative Pattern Casting (EPC) process and the Vacuum-Sealed Moulding (V-) process. Although quite different in terms of their requirement for patterns, they are similar in terms of their use of dry, unbonded sand as the moulding medium. Such a material can be rigidified to a high degree by applying vibration, of suitable frequency and amplitude, to the container so that the sand is compacted to near its minimum theoretical volume and maximum density. In the case of the EPC process, vibration alone may be sufficient to achieve the required mould rigidity for casting purposes. However, in the case of the V-process, vacuum assistance is required to complement the rigidity obtained by vibration to ensure that the mould retains its rigidity.

The application of a vacuum to a bed of sand has a fundamental effect [1]. By reducing the pressure of air in the spaces between the sand grains, the force of grain-to-grain contact is increased by the differential pressure between the vacuum and atmosphere, thus the internal friction is increased and flowability reduced. A vacuum may be established in a sand bed by two methods. In one method the external surface may be sealed and the air pumped out of the system, thereby reducing the pressure of the remaining air, this is the approach used in the V-process. Alternatively, the characteristic limited permeability of the sand may be utilised to generate a pressure drop throughout the sand bed by maintaining a continuous passage of air through from the top to the bottom of the mould, the method used in the EPC process.

The unbonded sand processes are described in more detail in the following sections.

## THE EVAPORATIVE PATTERN CASTING (EPC) PROCESS

The Evaporative Pattern Casting (EPC) process, also known as the expanded polystyrene moulding process, the evaporative foam process, the lost-foam process, or by such trade names as Full Mould, Policast, Replicast and Styrecast, has been known to the foundry industry since 1958 [2]. The application of dry, unbonded sand to the process is credited to Smith, whose US patent was granted in 1964 [3]. Since the expiration of this patent significant development work has taken place, mostly associated with the automotive industry. During the latter part of the 1980's considerable interest has been generated in the use of the process for the mass production of castings.

In the USA a recent survey [4] has predicted that the output of aluminium alloy castings by the process will increase to over 152,250 tonnes (150,000 tons) by 1993 and that for iron castings to 121,800 tonnes (120,000 tons). By the year 2000 these tonnages are expected to double. Although other materials are suitable, expanded polystyrene is used predominantly as the pattern material. For a detailed review of the process the reader is referred to papers published by the author [5-7].

### Process Outline

Although patterns can be made by machining blocks of expanded polystyrene, this approach is really only suitable for prototype and small quantity production. Patterns for mass production are produced by steam moulding in metal tooling using techniques developed from those used to produce expanded polystyrene packaging and insulation products. The production of patterns commences with granules of polystyrene which are pre-expanded to form beads. After ageing these beads are steam moulded to produce an expanded polystyrene pattern. The pattern, coated with a suitable refractory wash, is embedded in dry, unbonded sand which is vibrated to produce a rigid mould. On pouring, the molten metal displaces the polystyrene to produce the shape of the casting required. After shake-out, the casting requires minimal fettling because there is no mould joint line and the requirement for cores is eliminated. The moulding sand is entirely reclaimable, with cooling and classifying the only treatments required. The process offers benefits to both the casting producer and user, and the advantages claimed for the process are listed as follows:

- reduced capital equipment costs.
- reduced space requirements.
- conventional coremaking machines, dryers, moulding machines and mullers eliminated.
- skilled labour requirements reduced.
- process readily automated.
- reduced number of process variables.
- potentially lower tooling costs.
- reduced maintenance requirements.
- inexpensive, standardised moulding boxes.
- no sand binders.
- sand is reusable.
- shake-out is simplified.
- sand handling is simplified.
- cores are eliminated.
- cleaner and quieter working conditions.
- greater component design flexibility.
- improved casting yield.
- potential for reduced casting weight (no taper).
- improved casting quality.
- no mis-match, coreshifts, swells, run outs or dirt in mould.
- reduced fettling and machining.

The disadvantages associated with the process include the following:

- quality of casting determined by quality of pattern.
- surface finish determined by that of the pattern.

- multi-piece patterns may be necessary for complex shapes.
- patterns easily damaged.
- limitations on the minimum section thickness of pattern.
- patterns subject to shrinkage (but this is predictable).
- carbon pick-up in low carbon steels.

**Tooling Requirements**

Precision-machined tools are required for the process. Aluminium alloy is the most commonly used material for the tools because of its attractive features, which include: good thermal conductivity, good corrosion resistance, fair strength, light weight, and the ability to take a good surface finish on machining. The incorporation of loose pieces within the tooling may permit single piece patterns to be produced. However, for complex shapes, it is inevitable that the pattern will be produced from a number of moulded parts which are subsequently joined together [8]. The tooling must make provision for mould venting, steam injection, bead injection and pattern ejection. These factors are important for the attainment of uniform pattern density and minimum pattern distortion. The nature of the pattern production process is such that there is little wear in the tooling.

**Pattern Production**

The equipment required for the manufacture of patterns includes a bead pre-expander, a moulding machine and a steam-raising boiler. Facilities will also be required for the ageing and storage of pre-expanded beads, and for the ageing and storage of patterns. The pattern exerts a fundamental influence on the quality of the casting, so it is imperative that all stages in the production and preparation of the patterns be carefully controlled.

***Pre-Expansion.*** The granules of polystyrene, copolymerised with a low boiling point liquid such as pentane, are processed in batch pre-expanders in which heat is used to soften them. At a temperature of 80°C (145°F) a pentane vapour phase is produced and, as the temperature increases to 100°C (212°F), the vapour expands and this increases the diameter of the individual grains to produce polystyrene beads. After ejection from the pre-expander the beads are cooled and dried. They must also be aged or conditioned for a period of time, during which their internal pressure equilibrates with atmospheric pressure, this process takes several hours. The shelf life of the conditioned beads is about 24 hours.

***Pattern Moulding.*** The pre-expanded beads are converted into the final pattern shape by steam moulding them in aluminium-alloy moulding tools mounted on a press. Typically, the sequence of events will be: close press; pre-heat the mould with steam; inject the pre-expanded beads using a blow-fill gun; fuse the beads by injecting steam which softens the beads, promotes expansion due to the residual pentane, and finally causes fusion; water cool the mould and pattern; open press; and eject the pattern [9]. It is important to cool the pattern in the mould so that the internal pressure of the beads is reduced to a point where the part will not expand when it is removed from the mould.

***Bead Size and Pattern Density.*** The quality of a polystyrene pattern is a function of bead size and density. The best strength and surface finish are achieved in patterns of high density and small bead size, but this is not commensurate with good casting quality. Although gas evolution is a function of pouring temperature and pattern density, it is a prerequisite that the latter be low to minimise gas evolution. In general, pattern densities between 16 and 24 gm/litre (16 and 24 oz/ft$^3$) are used, depending on the part thickness and metal to be cast.

***Pattern Shrinkage.*** Expanded polystyrene patterns are subject to shrinkage after leaving the mould. This is due to the evaporation of residual water and loss of the expanding fluid from within the beads [8]. The extent of the shrinkage is a function of several parameters and may continue for a period of up to 30 days following manufacture. It is, however, predictable [10].

***Pattern Assembly.*** The assembly of multi-piece patterns and the attachment of gates, runners and feeders is usually accomplished with the aid of adhesives. They should not react with the polystyrene or produce any ash. It is essential that the adhesive should seal the joint in order to avoid the ingress of the pattern coating material [8]. Hot melt adhesives have been used successfully.

***Pattern Coating.*** The coating of the pattern with a refractory wash is an essential prerequisite to the successful production of castings by the process [11]. The refractory layer effectively acts as the mould cavity surface layer, providing the support that prevents unbonded sand entering the cavity as the metal replaces the polystyrene. The coating controls the rate at which gases, produced by the thermal degradation of the polystyrene pattern, are removed. Consequently, it should not be too thick or its permeability will be impaired. However, if it is too thin, sand burn-on and erosion may occur [8]. The preferred method of coating patterns is by dipping and the coating may be dried by the use of circulated warm air, at a temperature below 50°C (120°F) to avoid distortion.

***Alternative Pattern Material.*** A problem associated with the degradation of polystyrene patterns during the casting of irons is the formation of lustrous carbon defects in the surface of the casting. This is a consequence of the random scission mechanism by which polystyrene decomposes under the action of heat. An alternative foam material which has been used successfully in the production of castings is poly methyl methacrylate (PMMA) which decomposes by an unzipping mechanism [12]. Although this material generates greater volumes of gas during decomposition than does polystyrene, it does not produce lustrous carbon.

**Mould Production**

Mould production entails the investment of the coated pattern assembly in a dry, unbonded sand which is vibrated to obtain compaction.

***Moulding Sand.*** In the EPC process the casting surface quality is a function of the pattern surface quality and the pattern coating. The metal does not contact the moulding sand. For this reason sands are chosen to provide good compactability and high permeability. A sand with an AFS grain fineness number around 50 and with a size distribution predominantly on a single sieve would meet the requirements [8].

***Moulding Boxes.*** These are essentially open-top containers of rigid construction to withstand vibration. They may incorporate provisions for fluidising the sand or for vacuum suction.

***Pattern Investment.*** The pattern clusters may be positioned on a sand bed with additional sand introduced through a chute, in a way that does not damage the pattern coating, or embedded in a fluidised bed. The pattern may be placed by hand or be supported by jigs to prevent deformation.

***Vibration.*** Vibration of the sand around the pattern to obtain the optimum density and compaction of the mould is a critical feature in the successful production of castings by the process. The vibration characteristics are critical since the sand must be caused to flow into internal cavities and undercuts as well as be compacted completely [13]. The frequency, amplitude and intensity of vibration, the direction of movement imparted, the shape of the container and the point at which the vibration is applied all influence the quality of the castings produced [8].

Figure 4.1 shows the filling of a moulding box containing a polystyrene pattern with dry, unbonded sand.

Fig. 4.1 Photograph showing the filling of a moulding box containing a polystyrene pattern with dry, unbonded sand

**Mould Preparation for Casting**

Following vibration, the mould is prepared for casting, which may simply involve the location of a pouring basin above the sprue entrance. However, to prevent fluidisation of the sand during pouring, caused by the evolved gaseous pattern decomposition products, mould weights may be necessary, or the top of the box may be covered with a plastic sheet and a soft vacuum applied.

## Pouring, Gating and Feeding

It is necessary to cast at full flow from the very first moment to prevent even the smallest part of the sprue from burning instead of evaporating. In the event of combustion the sprue may collapse, with the result that the sprue becomes partly or completely blocked. The metal must be poured without interruption otherwise the pattern, and consequently the mould, may collapse [8].

The pouring rate is determined by the density of the pattern, the permeability of the sand, the metal being poured, and the number of gates employed. Gating is critical but, as yet, no firm guidelines are available. Research is being conducted to establish a better understanding of the way in which moulds fill [14].

## Vacuum Assistance

The application of vacuum suction to the mould prior to, and during, pouring can be advantageous. The vacuum assistance can be obtained by using a moulding box with a plenum chamber in its base so that air is drawn downwards from the open-topped mould to the plenum chamber, thus creating a pressure differential from top to bottom during casting [13]. The combination of vibration and vacuum assistance so improves the rigidity of the sand that the mould will not move under the effect of metallostatic pressure and the need for mould weighting is eliminated. Furthermore, the extraction of the gases generated during pattern degradation, at source and at the rate of generation, prevents fluidisation of the sand, fume and smoke emission, and the danger of metal blow-back.

## Casting Finishing

The absence of a mould joint line or the need for cores in the production of castings by the EPC process considerably reduces the requirement for finishing. This can be particularly advantageous in the production of castings in wear resistant alloys.

## Reclamation

The mould material is entirely reclaimable, although a small addition of new sand will be necessary to make up losses in the system. The only treatments of the sand required are cooling and removal of the fines. Both of these requirements can be achieved in fluidised bed, cooler-classifier equipment.

## Health and Safety Aspects

Expanded polystyrene has a low order of toxicity but is combustible. Pentane vapour is combustible when mixed with air in concentrations between 1.4 and 7.8% by volume [10]. Adequate ventilation is essential during the processing and storage of the material; sources of ignition should be eliminated and good standards of housekeeping and safety should be maintained. Some of the products of pattern decomposition are flammable and/or toxic and suitable arrangements must be made for their extraction at the metal pouring stage.

## Process Capabilities

***Alloy Range.*** The materials most commonly cast by the process are: aluminium-alloys, cast irons, carbon and low alloy steels, and wear and heat resistant alloys.

***Casting Size and Weight Range.*** The mass produced castings will be of a similar size and weight to those produced by green sand moulding. Typically, these will be in the weight range of 0.5 to 25 kg (1 to 55 lb). However, larger castings are produced by the process on a regular basis. Castings weighing several tonnes, produced from machined patterns, are used for one-off and prototype production.

***Production Volumes and Lead Times.*** The process is sufficiently flexible to cope with one off or one million off. However, for series production a reasonable order quantity is necessary to justify the tooling cost. The lead time in the production of tooling is no greater than that required for gravity dies or investment casting wax dies. Once the tooling is available in the foundry, the response time for the production of castings is dependent upon the pattern production sequence.

***Thin Section Capability.*** Sections of 3.50 mm (0.14 in) can be cast by the process.

***Cast Holes.*** Holes of $\phi$ 3 mm ($\phi$ 0.12 in) can be cast by the process.

***Detail Replication.*** Reasonable detail can be provided by the process but it should be noted that cast surfaces may show the bead structure of the pattern.

***Surface Finish.*** This is a function of the pattern surface quality and the pattern coating. Surface finish is at the better end of the sand casting range of 6.3 to 25 μm (250 to 1000 μin).

***Dimensional Accuracy.*** Although published information is not available, the process produces castings which are more accurate than the green sand process. Machining is still necessary for critical dimensions. Reductions in machining requirements arise principally because there is no mould joint line, and therefore no casting taper, and cores are not required to produce external or internal casting features.

***Typical Castings.*** Aluminium alloy castings for the automotive industry include: inlet manifolds, heat exchangers, cylinder heads and blocks. In cast irons: crankshafts, ventilated brake discs, exhaust manifolds, differential cases and diesel engine cylinder blocks. Non-automotive applications include: pipe fittings, valve castings and wear resistant castings. Examples of castings produced by the process, and the patterns from which they were made, are shown in Fig. 4.2.

## THE VACUUM-SEALED MOULDING PROCESS

The Vacuum-Sealed Moulding process, or V-process as it is commonly known, was developed in Japan in 1971 and introduced to Japanese foundrymen at their Spring meeting in 1972. By mid-1973 30 licenses for the process had been granted to Japanese foundries and by 1984 the number of licences granted had increased to 130 [15]. In Western Europe 37 licenses had been granted and in 1985 it was estimated that these licensees were producing 2100 tonnes (2058 tons) of castings/month compared with 9100 tonnes/month (8920 tons), representing about 2% of total national castings output, in Japan. Although it cannot be described as a major foundry process, only four UK foundries were using the process in 1988 [16], there are still developments being undertaken [17] which could lead to its wider use. For a detailed review of the process the reader is referred to a paper published by the author [18].

### Process Outline

In the process the pattern assembly is mounted on a hollow pattern box which is connected to a vacuum pump. Vents in the pattern/board assembly enable air to be drawn through the assembly by the pump. A thin plastic sheet is heated in a position directly above the pattern until it softens, whereupon it is vacuum-formed to the contour of the pattern assembly by vacuum suction applied to the pattern box. At this stage a refractory mould coat may be sprayed over the plastic sheet and dried before the specially constructed moulding box is located onto the pattern box assembly. The moulding box is filled with dry, unbonded sand which is vibrated to optimise packing and bulk density. A second plastic sheet is used to seal the top of the moulding box. Vacuum suction is applied to the sand through the specially constructed moulding box to hold the sand in a rigid and compact state. At this point the vacuum suction is released from the pattern box to enable the mould to be separated from the pattern. Vacuum suction is maintained in the mould.

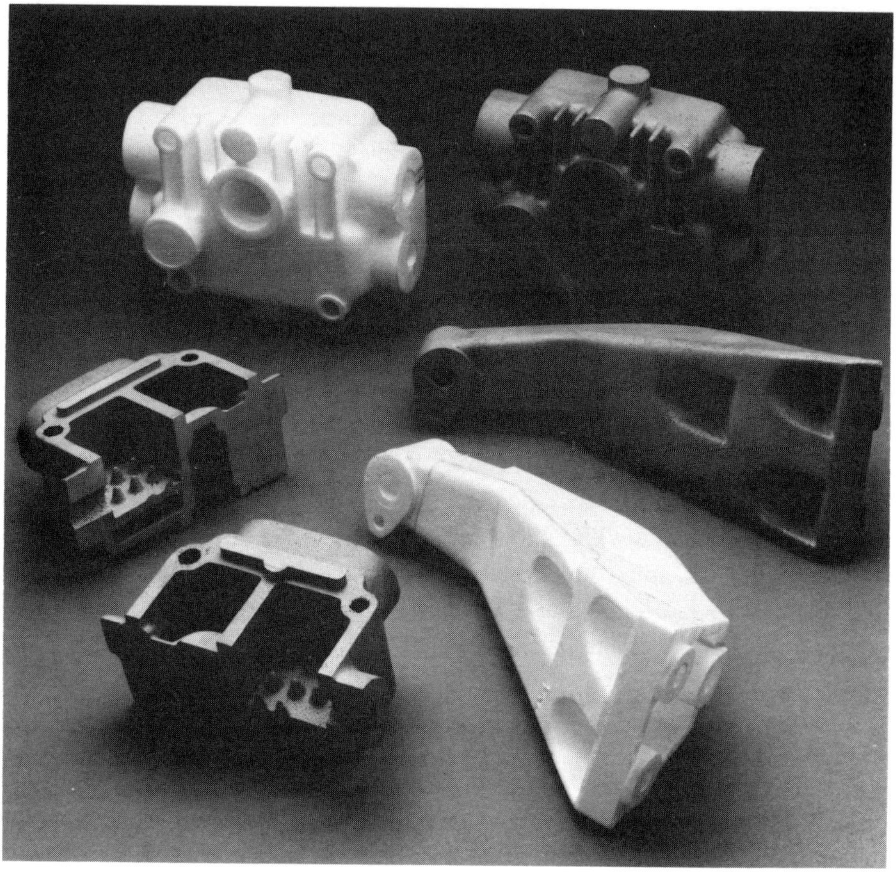

Fig. 4.2 Photograph showing polystyrene patterns and the castings produced from them

Two half moulds produced in this manner are located together, after coring if required, and are transferred to the pouring station. Vacuum suction is maintained in the mould at all times. After pouring, and when the casting has solidified, the vacuum is released. Following a period of cooling the sand and casting are removed from the box; the sand is recycled for future use and the casting is processed through the fettling shop.

A considerable number of advantages are claimed for the process, and include [15, 19]:

- existing patterns can be converted.
- low pattern costs and little pattern wear.
- pattern taper reduced or eliminated.
- consistently hard moulds.
- good dimensional accuracy.
- good surface finish.
- improved mould filling characteristics.
- more favourable feeding characteristics.
- low casting stresses and susceptibility to cracking.
- very low fettling costs.
- reduced casting scrap.
- reduced machining costs.
- environmentally favourable process (reduced fume, noise and waste).
- no binders or water.
- sand fully recyclable (1-2% loss).
- versatility in respect of casting alloy, section and size.
- ease of pattern separation from the mould.
- reduced moulding box maintenance.

Against these advantages, the following disadvantages must be balanced:

- high capital costs at low mould outputs.
- increased standard for plant and equipment maintenance.
- high licensing fees.
- high energy consumption.
- strict quality control for raw materials (film) essential.
- close process control necessary.
- expensive, specially constructed moulding boxes required.
- mould coating required.
- long cooling times - may require a large stock of moulding boxes.
- some configurations, e.g. narrow recesses and pockets, do not lend themselves to the process.

- carbon pick-up from the film may occur in low carbon steel.
- casting skin may be decarbonised and graphite form degraded in iron castings.

**Tooling Requirements**

Good quality patterns, of a standard similar to that required for high pressure green sand or cold set moulding, are required for the process and may be produced in wood, plastic or metal. The patterns are not subjected to wear or mechanical damage during moulding because the plastic film protects the pattern surface from the abrasive action of the sand and force is not required to produce the mould or separate the pattern from the mould. The patterns should be produced to a high standard of surface finish, otherwise poor surface detail will be readily replicated on the casting [20].

It is essential to provide adequate venting of the pattern board, and in some cases the pattern, to ensure that the plastic film is drawn faithfully over the detail of the pattern. When the mould has been produced and the vacuum to the pattern box released there is a slight compression of the mould which produces a small, uniform clearance that improves pattern stripping [21]. This may permit patterns with zero taper to be used, although 1° of taper is more usual, which is significantly less than that required for bonded sand moulds and reduces machining costs [20].

**Mould Materials**

***Sand.*** In any refractory aggregate moulding process, fine grains will improve casting surface finish at the expense of mould permeability. In the V-process the requirement for high permeability is obviated by the absence of gas-generating material in the mould and by the fact that the air in the cavity can be removed by vacuum suction during the pouring stage. The vacuum suction can promote metal penetration, so that fine grained sands were preferred to reduce this tendency. Although fine grained sands were originally specified for the process, this constraint has largely been eliminated by the development of suitable mould coatings [22]. Sand grain size, shape and distribution are important in the compaction of the sand, in determining the amount of air which must be drawn from the mould by the vacuum pump, and in determining the dimensional accuracy of the casting [22-24].

***Mould Coatings.*** Mould coatings are applied to the sand side of the plastic film to overcome such problems as: metal penetration, pin-hole porosity, sand inclusions and mould collapse [25]. The coating must not react with the plastic film and should dry quickly and completely.

***Plastic Film.*** Ethylene vinyl acetate (EVA) is the preferred plastic film for the V-process. It is a thermo-plastic co-polymer of polyethylene with a vinyl acetate content of between 14 and 18% [26]. The characteristics required of the film include the following [22, 23, 27]:

- ability to stretch and deform plastically.
- ability to adhere to the pattern during forming but not stick to the pattern during pattern stripping from the mould.
- the film must have suitable surface characteristics to permit the adhesion of the mould coat.
- the film should generate minimum gas on contact with the metal.
- any gases generated should be non-toxic.
- it should be inexpensive.
- the film should be free from defects such as bubbles, tears and pin-holes.
- shape defects, wrinkles and overlaps, should not develop during forming.
- the film should not retain elasticity after forming, otherwise it may shrink away from the pattern as it cools.

A low-density polyethylene film, such as commercial packing film, may be used to seal the top of the mould [22, 28].

**Mould Production**

The production of the mould involves the following stages: film heating and forming; moulding box positioning, sand filling and compaction; mould sealing; and vacuum application.

***Film Heating and Forming.*** The heating and forming of the plastic film to the contour of the pattern equipment is clearly a critical stage in the production of a V-process mould. The film must be heated uniformly over its entire area to a constant temperature before heat, and consequently formability, is lost [23]. A film thickness of 0.05 to 0.10 mm (0.002 to 0.004 in) is generally recommended but after forming the thickness may reduce to 0.006 mm (0.0002 in). Warm air, gas or electrical radiation heaters, or contact plate heating of the film, have all been practised, with close control of the heating cycle recommended to ensure consistency in operation.

***Moulding Boxes.*** In addition to providing a rigid frame to contain the sand, the V-process moulding box must also transmit the vacuum suction which maintains the mould in a rigid state. This rigidity is provided by the difference between atmospheric pressure and the reduced pressure within the mould. It is essential that the box construction ensures that the vacuum suction is applied to all parts of the mould. This can be achieved by using hollow wall construction with screened mesh openings and/or mesh-wrapped perforated tubes which span the box width.

***Sand Compaction.*** For a given sand, the higher the bulk density the higher will be the strength and hardness of the mould [27]. Precompaction of the sand by vibration is important to ensure that a high bulk density is obtained before the vacuum suction is applied. Variations in the frequency, amplitude and duration of the vibration affect the degree of compaction obtained and must, therefore, be carefully controlled.

***Vacuum Requirements.*** The vacuum must be maintained within the mould from the time that it is poured until the resultant casting has solidified, with the most critical period occurring during pouring. The vacuum plant must first remove the air in the mould which, even in a well-compacted mould, can represent 30% of the mould volume. This is achieved within seconds and thereafter the plant copes with a small amount of air leakage until pouring commences. In order that the mould should not collapse during pouring, it is essential that air be present in the mould cavity throughout pouring to provide the necessary pressure differential. This is ensured by the provision of a vent channel called a "communicating hole". The loss of plastic film during pouring allows air to be drawn through the sand by the vacuum pump. The vacuum plant must have the capacity to handle the tremendous increase in the volume of air to be removed at this stage. The recommended level of vacuum is around 0.5 bar (7 lbf/in$^2$) during pouring. Figure 4.3 shows a half mould produced by the process.

## Gating, Feeding and Pouring

Fast pouring times have been recommended to minimise the danger of loose sand separating from the mould, a problem largely overcome by the use of a mould coating, and to reduce the demands on the vacuum plant caused by the inrush of air through the "communicating hole". However, the fast pouring time should be accomplished with quiescent filling using a non-pressurised gating system [27]. Feeding provision is necessary and should be determined according to accepted rules. It is possible that feeding may be unnecessary for certain compositions of grey cast iron as a rigid mould is assured.

Fig. 4.3 Photograph showing a mould half produced by the V-process (vertical joint line)

**Reclamation**

The mould material is recyclable, with losses of only 1 or 2% likely. An effective sand cooling system is required to ensure that the temperature of the recycled sand is below 40°C (105°F) to prevent undesirable plastic film softening.

**Health and Safety Aspects**

The inert nature of the mould material eliminates the problem of fume emission. However, the generation of fine siliceous dust during sand handling necessitates the provision of efficient dust extraction facilities.

**Process Capabilities**

***Alloy Range.*** The process has been used successfully for the production of castings in aluminium- and copper-base alloys, cast irons, and carbon and low alloy steels.

***Casting Size and Weight Range.*** The process is suited to the production of castings in the weight range of 100 to 2500 kg (225 to 5500 lb). The process is particularly suited to the production of castings with quite large areas and shallow depth.

***Production Volumes and Lead Times.*** Lead times are no longer than those for conventional sand casting processes. The process is suited to small batch, jobbing production, with outputs typically between 5 and 10 moulds/hour. However, for longer production runs, mechanisation can provide outputs of between 50 and 60 moulds/hour.

***Thin Section Capabilities.*** This is good in comparison with other sand casting processes and sections as thin as 3 mm (0.12 in) are possible.

***Detail Replication.*** Detail replication is good as the sand effectively "sets" in contact with the pattern. The process has found use in the production of decorative panels and cast tooling.

***Surface Finish.*** A good standard of surface finish is obtained in castings produced by this process. The nature of the process is such that surface finish values in the range 3.2 to 12.5 μm (125 to 500 μin) can be expected.

***Dimensional Accuracy.*** Although detailed information is not available in the literature, there is evidence that the process produces castings which are more accurate than those obtained from the green sand or cold set processes.

***Typical Castings.*** Examples of castings produced by the process include those for the machine tool, textile machinery, printing machinery, earth-moving and valve industries. Specific examples include: cast iron bath tubs, piano frames, printing machinery side frames, milling machine saddles, fork-lift truck counterweights, valve castings, etc.

## References


1. Bish, D. A. and Hubbard, P. A. (1982). Application of ceramic coatings to expendable patterns. *Proceedings of the SCRATA Annual Conference*, Paper No. 17, 6 pp.

2. Shroyer, H. F. (1958). Cavityless casting mold and method of making same. *US Patent 2,830,343*.

3. Smith, T. R. (1964). Method of Casting. *US Patent 3,157,924*.

4. Anon. (1990). EPC research: creating a competitive US market. *Modern Casting*, **80**, 6, 61-63.

5. Clegg, A. J. (1978). The full-mould process - a review. Part I: production and preparation of patterns. *Foundry Trade Journal*, **145**, 3141, 149-158, 160.

6. Clegg, A. J. (1978). The full-mould process - a review. Part II: production of castings. *Ibid*, **145**, 3143, 393-403.

7. Clegg, A. J. (1986). Expanded-polystyrene moulding - a status report. *Foundry Trade Journal International*, **9**, 30, 51-61, 69.

8. Goria, C. A., Del Gaudio, G., Caironi, G., Silva, G. and Selli, M. (1983). Moulding of iron castings with evaporative polystyrene foam patterns immersed in loose sand: prospects and problems. *Metallurgical Science and Technology*, **1**, 1, 14-31.

9. Bailey, R. (1982). Understanding the evaporative pattern casting process (EPC). *Modern Casting*, **72**, 4, 54-61.

10. Immel, R. H. (1979). Expandable polystyrene and its processing into patterns for the evaporative casting process. *Trans AFS*, **87**, 545-550.

11. Del Gaudio, G., Serramoglia, G., Caironi, G. and Tosi, G. (1985). Aspects concerning the role of coatings in the production of castings by the Policast process. *Metallurgical Science and Technology*, **3**, 3, 76-86.

12. Moll, N. and Johnson, D. (1986). A new moldable foam for casting ferrous materials without lustrous carbon defects. *Proceedings of the Advanced Casting Technology Conference*, 175-185. Published by ASM International, 1987.

13. Brown, J. R. (1982). The Replicast process. *Foseco Foundry Practice,* 205, 3-5.

14. Graham, N. K. and Irvine, K. R. (1990). Computerised metal flow analysis in the lost foam process. *Paper presented to the Annual Conference of the Institute of British Foundrymen,* 20 pp.

15. Engels, G. and Schneider, G. (1986). Current situation and development trends of the vacuum moulding process. *Casting Plant + Technology*, **2**, 2, 12-20.

16. Anon. (1988). Major investment in Black Country foundry. *Foundry Trade Journal*, **163**, 3373, 551.

17. Campbell, J. (1988). Some recent developments in the aluminium casting industry. *Cast Metals,* **1**, 3, 156-160.

18. Clegg, A. J. (1985). The V-process - review and current status. *Foundry Trade Journal*, **158**, 3307, 472-477, 486.

19. Mizuno, T. (1977). Current status of V-process moulding. *Proceedings of the Seminar on Engineering Equipment for Foundries and Advanced Methods of Producing Such Equipment*, 323-334. Published by Pergamon Press, 1979.

20. Troland, W. W. (1982). The new V-process is 10 years old. *Casting Engineering/Foundry World,* **14**, 1, 42-47.

21. Dreger, D. R. (1978). Smooth no-draft castings. *Machine Design*, **50**, 12, 63-65.

22. Green, G. A. (1982). Superior castings and improved environment from the V-process. *Castings,* **28**, 5/6, 30-36.

23. May, R. P. (1979). Equipment and materials for V-process moulding. *The British Foundryman*, **72**, 4, xxxviii-xivii.

24. Gaindhar, J. L., Jain, C. K. and Subbarathnamatah, K. (1985). Effect of sand variables on dimensional accuracy of Al-4% Cu castings produced by V-process. *Trans AFS,* **93,** 303-306.

25. Kasai, H. (1980). High production of thin-wall castings by the V-process. *Ibid*, **84**, 535-540.

26. Enderle, R. J. (1979). Introduction of vacuum molding at Buckeye Steel. *Ibid*, **87**, 59-65.

27. Kono, R. and Miura, T. (1974). Fundamental techniques of the vacuum-sealed moulding process. *The British Foundryman,* **67**, 12, 319-325.

28. Salmons, J. A. (1982). Some practical aspects of the work at Crown Foundry Ltd. *Ibid*, **75**, 12, 240.

## CHAPTER 5

# The Plaster Mould Casting Process

The plaster mould casting process is a specialised process used for the production of non-ferrous alloy castings with smoother surfaces, more finely reproduced detail and greater dimensional accuracy than can be obtained from sand moulds, including shell moulds, or gravity dies (permanent moulds). Although there are plaster-based investment materials which are used with expendable patterns, they are best considered as a variation on the investment casting process [1]. The true plaster mould casting process is based on the use of a permanent pattern. In the commercial production of castings three variants of the process are used:

The Conventional Plaster Mould Casting Process
The Foamed Plaster Process
The Antioch Process

In the conventional process, plaster of Paris is mixed with water to produce a slurry which is poured over a permanent pattern contained within a moulding box. Upon setting a rigid mould is produced which, after pattern stripping, is dried at an elevated temperature to remove free and chemically combined water before the metal is cast into the mould. This method produces a strong, dense, but inherently impermeable mould. Hence, metal casting must be conducted using vacuum and/or pressure assistance to ensure complete filling of the mould by the metal. The insulating nature of the plaster improves the fluid life of the metal, which aids mould filling and thin section production.

The foamed plaster variant produces permeable moulds through the incorporation of air into the slurry at the mixing stage. It is possible to produce a mould in which 50% of the volume consists of air bubbles. This has the desired effect of increasing permeability from 1 or 2 units to between 15 and 30 units, as measured by the AFS permeability test [2]. It also provides a more economical use of the plaster. However, the inherent strength of the mould is reduced and if complex shapes are to be produced then flexible patterns, traditionally rubber, may be required.

The Antioch process has, as its special feature, the requirement to process the moulds in a steam autoclave. This produces a unique granular structure which provides mould permeability. The moulds produced by the Antioch process are denser than foamed plaster moulds and weaker than conventional plaster moulds. However, this lower strength does not manifest itself until after mould production, which enables permanent patterns to be used.

The advantages and disadvantages of the plaster mould casting process are summarised below [2-4].

Advantages:

- the ability to produce complex shapes.
- the ability to produce thin section castings.
- the excellent replication of pattern detail.
- the ability to produce castings which are dimensionally accurate.
- the ability to produce castings with good surface finish.
- the minimisation of residual stresses and distortion in castings.

Disadvantages:

- poor productivity due to lengthy processing procedures.
- the need for multiple patterns to improve moulding productivity.
- the requirement for close control of the production process.
- the need for special procedures to overcome the problems of poor mould permeability.
- the possibility of impaired mechanical properties arising as a result of slow cooling of the casting.
- the mould materials are not reclaimable.

## Mould Materials

The principal materials used in the production of a mould are plaster and water. However, other materials are used to develop required mould characteristics and there are a number of process variables which must be carefully controlled. The chemistry of the process must be considered as it exerts an influence on both mould processing and casting quality. Gypsum plaster consists of a mixture of three related chemicals:

calcium sulphate dihydrate ($CaSO_4.2H_2O$)
calcium sulphate hemi-hydrate ($CaSO_4.0.5H_2O$)
calcium sulphate anhydrite ($CaSO_4$)

The different proportions of these three chemicals determine the plaster's properties [5]. Plaster of Paris is pure hemi-hydrate and is produced by calcining the dihydrate at a temperature of 128°C (262°F). The hemi-hydrate is available in two principal types, referred to as α and β [5]. The α form is produced under pressure in a humid atmosphere, which results in an even crystalline structure. When the hemi-hydrate is produced at atmospheric pressure the water is given off as steam, which disrupts the crystalline structure, resulting in the fragmentary crystals of β hemi-hydrate. The principal differences in their practical characteristics are that β requires more water to produce a plaster and the resulting plaster will set more quickly and be weaker.

The chemical change of calcination is reversible by simply adding water to the hemi-hydrate and this process is known as hydration. During hydration crystals of dihydrate are produced. Initially these crystals will be in suspension, floating freely in the liquid of the plaster slurry. However, with time these crystals will grow to interfere with each other and join up to form a crystal matrix, this is known as the initial set. The process of hydration continues until all the hemi-hydrate is converted to the hydrate, this is known as the final set [5]. In theory, the amount of water required to hydrate pure plaster of Paris is 18.6% by volume of the plaster [3]. However, in practice, more water than is actually necessary for hydration is required so that hydration will be complete before the plaster has actually dried out.

The initial set does not occur simultaneously throughout the plaster slurry; there is, however, a more or less gradual process of setting. The setting process can be manipulated by adding accelerators or retarders to the mixture. Hydrated plaster is an effective accelerator and an unwelcome one if its presence is due to improper cleaning of mixing equipment.

Heating of the hydrated plaster above 163°C (325°F) removes all of the water of crystallisation and produces the anhydrite. At a temperature of 1195°C (2183°F) the plaster decomposes to calcium oxide [2].

***Water.*** Water purity is important as the presence of organic materials causes variations in setting time and soluble salts can migrate to the mould surface during mould drying, causing efflorescence and hard spots which might impair surface smoothness [6]. Providing that the water supply is potable it will probably be suitable for the process. The temperature of the water should be maintained within the range 20 to 38°C (68 to 100°F) to ensure maximum solubility of the plaster. It is advisable to use a heated water tank to provide a controlled water temperature in order to minimise the variations in setting time caused by variations in water temperature.

Variations in the water:plaster ratio will affect the mould strength and performance [3]. It is essential that both materials should be weighed accurately before mixing. Although in theory the volume of water required for hydration is only 18.6% of the plaster volume, in practice considerably more water is required to provide slurry fluidity. A typical mix ratio of plaster to water on a weight basis is 1:1.1 [7].

***Plaster Additives.*** Proprietary plasters usually contain additives to enhance such mould properties as: green and dry strength, permeability and refractoriness [2]. The additives used for these purposes include: cement, clay, silica, fibre glass, and talc. Formulations for the Antioch process may contain less than 50% of plaster.

***Soaking.*** It is recommended that dry additions be allowed to soak in the water for 2 to 3 minutes before mixing commences [6]. This enables the water to wet the plaster thoroughly and aids its dispersion during subsequent mixing. However, soaking is not necessary for the foamed plaster mix as a wetting agent is used.

***Mixing.*** Mixing is an important stage in mould production because it influences such properties as mould strength, hardness and permeability [6]. Variations in mixing procedure have a greater effect on mould quality than any other stage in mould production. For the conventional process the plaster can be mixed in a bucket type container using a propeller positioned at the end of a shaft [2]. The propeller should be situated low in the container so that air entrainment is minimised. Over-mixing should be avoided as it will hasten setting and may increase expansion upon setting [8]. Under-mixing will result in watering-out and weak moulds.

To produce foamed plaster a foaming agent, such as alkyl aryl sulphonate, is added either to the dry ingredients or the slurry [2]. The mixer design must ensure that air is beaten into the slurry as the objective is to produce a mixture consisting of 50% by volume of air. The air bubbles should be no larger than $\phi$ 0.25 mm (0.01 in) and they should remain under the surface skin of the plaster. Mixer designs based on perforated discs or double-ply rubber discs are recommended.

## Tooling Requirements

Pattern equipment for the process may be either rigid (permanent) or flexible. Permanent patterns are usually associated with match-plate mould production so that cope and drag plated patterns, incorporating the necessary gating and feeding system, must be produced. Wood, plastic or metal may be used, with qualifications, as the pattern material. However, the principal requirement in pattern production is precision as this will ultimately determine the accuracy and surface quality of the casting.

***Wood.*** Wood is cheap and easily shaped. However, tool and grain markings can cause problems at the pattern withdrawal stage as the expansion of the plaster during setting can cause it to lock into pattern imperfections. The surface of the wood must be sealed but, even so, intermittent contact with the water-based plaster slurry inevitably impairs the dimensional stability of the pattern. Wood should only be considered for small quantity production or for cast-to-shape production of permanent patterns in metal or epoxy resin.

***Plastics.*** Patterns produced using epoxy resin can be very accurate when properly made. However, care must be taken to avoid distortion in heavy sections and thin sections may be brittle [1]. These problems can be overcome by using metal parts for the thin sections and by lightening out the heavy sections.

***Metals.*** Patterns produced in metal are preferred for production runs because they have excellent dimensional stability and can be produced to a high standard of surface finish [1]. Aluminium alloy is a popular material because it is light, easy to machine and has good corrosion resistance. Brass is also acceptable as it is easily worked, has the necessary hardness, corrosion resistance and dimensional stability [6]. The material will also permit soldering to seal joints which can be an advantage. Ferrous metals, other than austenitic stainless steels, cannot be used as they would corrode in service. Care must also be taken when dissimilar metals are used in pattern construction. If the pattern surface is not sealed, electrolysis will occur when the plaster slurry contacts the pattern. This will produce bubbles in the plaster as well as cause corrosion of the patterns [1]. Metal patterns are not prone to wear when used with this process and the absence of pressure or other forces during mould production means that pattern structure can be relatively light.

Patterns must incorporate an allowance for contraction and this will vary with the size, configuration and section thickness of the casting [3]. For aluminium alloys the contraction allowance will generally lie between 1 and 1.5%. Taper is required on pattern surfaces perpendicular to the mould joint plane. Exceptionally this may be as low as 0.25°, but is more likely to be between 0.5 and 2.0°.

Cores or loose pieces may be used to produce undercuts or other features, with preference given to loose pieces where practicable, as they tend to provide greater accuracy in the casting [1].

The comments on permanent patterns apply equally to permanent core boxes.

***Flexible Patterns.*** The use of flexible patterns is a common feature in the plaster mould process [1]. Although commonly referred to as rubber, because traditionally a polysulphide rubber was used, urethane is now preferred because it produces tougher and more durable patterns. Flexible patterns are used primarily with foamed plaster moulds to eliminate the breakage associated with stripping the weak mould material from a permanent pattern. There is some danger of distortion of the flexible pattern due to the hydrostatic pressure of the plaster slurry while the mould is being made. The life of the pattern may be limited if it must be flexed too much to separate it from the mould. This type of pattern lends itself to the production of shapes with complex undercuts, an example being the production of the tread pattern in tyre moulds. The advantages of flexible urethanes may also be extended to coremaking.

## Mould and Core Production

Patterns may be coated with light oils or waxes, for example paraffin and/or stearic acid, to aid separation from the mould [9]. However, these parting agents should be applied sparingly as they may impair the hardness of the mould surface. The slurry should be carefully introduced into the moulding flask to avoid air entrapment at the pattern surface. The fluidity of the slurry, which resembles heavy cream in consistency, allows it to flow around the pattern to produce moulds of uniform density with excellent finish and remarkably faithful reproduction of the pattern detail [1]. This latter feature is promoted because the plaster expands slightly as it sets [10]. Vibration may be applied to the pattern plate to assist in the removal of air bubbles from the pattern surface [6].

Plaster setting times may vary according to practice, but typically are between 10 and 20 minutes [1, 9]. Once established, they should be maintained to ensure consistency and maximum detail in the moulds. The intimate contact between mould and pattern requires that pattern stripping be conducted with care. A mechanical stripping device may be used to minimise damage to the mould at this stage. Compressed air may be injected through the pattern to the pattern/mould interface to aid separation. In some cases air is injected through vents in the back of the mould, where it drives moisture to the pattern/mould interface. This is termed "sweating" and aids pattern separation [9]. It is possible, with skill, to repair minor mould defects at this stage.

## Mould Drying

The mould drying stage is critical in the production of high quality moulds and the resultant castings. Careful selection of equipment having effective process controllers, especially for the time and temperature of drying, is an essential requirement. The recommendations for mould drying are similar for each of the process variants. The moulds should be dried as quickly as possible after stripping from the pattern to ensure that maximum physical properties are developed in the mould material.

A forced air circulation oven with adequate ventilation and thermostatic control is recommended [7]. Good circulation is necessary to ensure that even heating is obtained and the moisture laden air must be vented and replaced with drier air for the drying operation to be successful. Drying temperatures within the range of 200 to 260°C (390 to 500°F) are quoted as acceptable. However, to ensure consistent quality and reproducibility of mould dimensions, a specific temperature should be determined to suit production requirements and maintained within ± 6°C (± 10°F) [2]. The drying time depends primarily on the thickness of the mould material which, in turn, determines the time taken for the heat to penetrate to the centre of the mass. To determine this it will be necessary to conduct experiments in which thermocouples are inserted into the mould to determine the time taken for the centre of the mass to reach a specified temperature. It may be necessary to dry moulds for periods of time of up to 72 hours [10].

It is essential that the drying process removes both the free and combined water if gas defects in the casting are to be avoided. This explains why a minimum temperature of 200°C (390°F) is recommended. Although drying can be accomplished more quickly at higher temperatures, these may cause shrinkage and cracking of the mould. An allotropic change, associated with a volume change, occurs in the plaster at a temperature around 380°C (716°F).

***Antioch Process Variation.*** After setting, the moulds/cores are stripped from the pattern and steam autoclaved for 8 to 10 hours at a pressure of 2 bar (30 psi). During this time partial dehydration occurs and recrystallisation produces a granular structure rather than the usual acicular structure. After the autoclave treatment the moulds may be dipped in water and then allowed to stand for 10 to 14 hours to ensure that granulation is complete. The granular effect is not usually apparent at the surface of the mould but the mould develops improved permeability, typically in the range 25 to 40 AFS [11]. The moulds must be oven dried to remove all excess and combined water.

Figure 5.1 shows plaster moulds for the production of aluminium alloy compressor wheel castings.

## Mould Preparation for Casting

After drying, moulds are fitted with cores, if required, and joined ready for pouring. Across-the-joint location is usually achieved within the mould rather than by moulding box pin and bush. This is because differential expansion, between mould and box, occurs during mould baking and this causes the mould to loosen within the box [1]. Double-ended conical pins which sit in prints moulded in each mould half are an effective means of location. As the plaster is hygroscopic after drying, the moulds should be maintained at a temperature above 65°C (150°F) to minimise moisture pick-up from the atmosphere.

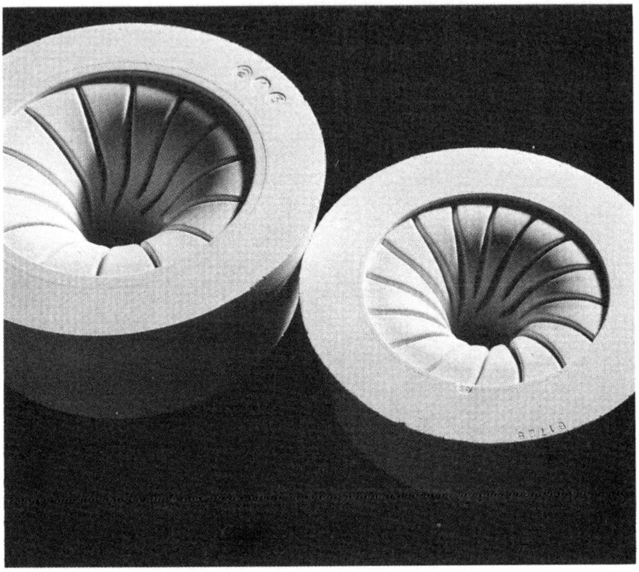

Fig. 5.1 Photograph of plaster moulds for the production of aluminium alloy compressor wheel castings

Vacuum and/or pressure assisted filling of the mould is essential for dense plaster moulds which have poor permeability [2, 4, 7]. Although casting in a vacuum chamber would be the ideal, soft vacuum, in the form of suction applied to the top and bottom surfaces of the mould, is acceptable. To improve efficiency the side surfaces of the mould can be sealed with a shrink-wrap, plastic film [9]. Pressure assisted filling of the mould can be accomplished by introducing the metal under a compressed air pressure head of up to 1 bar (15 psi). This pressure head can be maintained during solidification of the metal, thus providing pressure assisted feeding and, consequently, a sounder casting.

## Gating and Feeding

In general, the normal principles of gating and feeding apply to the production of castings by the plaster mould process. The insulating nature of the plaster promotes good fluid flow and aids the filling of thin sections. Hence, ingate sizes can be smaller than for equivalent sand castings [1]. For the production of high-integrity aluminium alloy castings the use of unpressurised gating systems is recommended [4]. Slow cooling may promote shrinkage in isolated sections and coarse grain structures. Directional solidification can be induced by the incorporation of metal chills in the mould [2, 3].

## Casting Finishing

After the castings have solidified the plaster mould material is readily broken down but, to avoid damage to the castings, a high pressure water jet may be used for blasting [1]. Otherwise, finishing is by conventional means.

## Reclamation

Although it is technically feasible to reclaim the plaster, this is not normally practised. Founders prefer the assurance of quality provided by new plaster supplies.

## Health and Safety Aspects

Inhalation of the fine air-borne dust associated with initial mixing of the plaster and knock-out of the moulds may pose a health hazard and should always be avoided. Protective face masks and well ventilated working areas are recommended. Water may be used at the knock-out area to suppress dust.

## Process Capabilities

***Alloy Range.*** The process is used predominantly for the production of castings in aluminium-base alloys. There is some production of magnesium- and zinc-base alloys by the process, although in the case of the latter this may be for proving a design which will subsequently be produced by diecasting [10]. Copper-base alloys may also be produced by the process, although the pouring temperature of these alloys may be close to the disassociation temperature of the plaster. Aluminium bronze, manganese bronze and 60:40 brass castings have been produced by the process [6].

***Casting Size and Weight Range.*** Castings weighing from as little as 50 gm (2 oz) to 10 kg (25 lb) are routinely cast and this probably represents the weight range of about 90% of the production output in aluminium alloys. Aluminium alloy castings up to 50 kg (110 lb) are produced and, exceptionally, a casting weighing 1815 kg (4000 lb) has been recorded [2].

***Production Volumes and Lead Times.*** Although precision patterns are required, lead times are shorter than those for investment casting or pressure diecasting and tooling costs modest by comparison. The process is most suited for small quantities but can be used for larger quantities, especially when duplicate patterns are used.

***Thin Section Capability.*** It is claimed that sections as thin as 0.25 mm (0.01 in) can be cast. However, more typical is the claim that section thicknesses of 1.5 mm (0.06 in), 1.75 mm (0.07 in), and 2.25 mm (0.09 in) can routinely be cast in zinc-, aluminium- and copper-base alloys respectively. For premium quality castings a section thickness of 1 mm (0.04 in) can be achieved for each of the alloy groups [4].

***Detail Replication.*** Although not easy to quantify, detail replication is excellent due to the fluid characteristics of the plaster, which ensures good contact with the pattern, and the expansion of the plaster on setting, which improves replication of pattern detail. The replication ability of the process is demonstrated by the fact that a finger print in the pattern coating will be replicated on the casting.

***Surface Finish.*** The surface quality of castings produced by any process tends to decrease with increasing metal pouring temperature. Hence, the values will be higher for copper-base alloys than for aluminium-base alloys. Generally, castings can be produced with surface roughness values within the range of 0.8 to 3.2 μm (25 to 125 μin).

***Cast Holes.*** Due to the relatively poor strength of the plaster mould material, the minimum hole or cut-out size which can safely be cast is $\phi$ 6 mm (0.25 in). It is recommended that holes be cast with a machining allowance of 0.4 to 1.5 mm (0.016 to 0.06 in) if precision is a requirement.

***Dimensional Accuracy.*** Attainable accuracy, in relation to size of dimension, is shown in Table 5.1. The figures represent the very best practice and do not include an allowance of ± 0.375 mm (0.015 in) for dimensions measured across the mould joint line.

Table 5.1 *Attainable Dimensional Accuracy in Castings Produced by the Plaster Mould Casting Process*

| Dimension Size | | Tolerance | |
|---|---|---|---|
| mm | in | mm | in |
| up to 75 | up to 1 | ± 0.175 | ± 0.007 |
| 25 to 75 | 1 to 3 | ± 0.250 | ± 0.010 |
| 75 to 150 | 3 to 6 | ± 0.375 | ± 0.015 |
| 150 to 300 | 6 to 12 | ± 0.500 | ± 0.020 |
| 300 to 600 | 12 to 24 | ± 0.625 | ± 0.025 |

***Typical Castings.*** Typical of the output of aluminium alloy castings by the process are bladed rotating castings for pumps and fluid transmission units [1]. Castings are also produced for the aerospace industry, fuel system components, fuselage components, etc; for business machines and computers; electronics and radar (waveguides); medical equipment; scientific instruments; optical equipment; building equipment; and engineering industries [1, 4].

Figure 5.2 shows a selection of typical castings produced by the process.

Fig. 5.2 Photograph showing a selection of aluminium alloy castings produced by the plaster mould casting process

## References


1. Petch, W. (1970). Plaster mold casting. *SME Technical Paper CM70-814*, 13 pp.

2. Nelson, C. D. (1988). Plaster molding. In: *Metals Handbook, Vol. 15 Casting*, 242-247. Published by ASM.

3. Shiota, T. and Ninomiya N. (1978). Precision casting of impellers. *SEAISI Quarterly*, **7**, 2, 53-61.

4. *The Plaster Mold Casting Handbook.* Published by The Plaster Mold Casting Association, Clifton, NJ, USA, 1983.

5. Coburn, A., Dudley, E. and Spence, R. (1989). *Gypsum Plaster, Its Manufacture and Use.* Intermediate Technology Publications, London.

6. Young M. K. (1975). Plaster and gypsum cements in the pattern shop and foundry. *Trans AFS*, **83**, 109-112.

7. *Herculite Foundry Plaster.* Product brochure published by Foseco International Ltd.

8. Anon. (1972). Plaster mold casting. In: *Metals Handbook, Vol. 5B Melting and Casting*, 222-231. Published by ASM.

9. Kanicki, D. (1978). Plaster molding for precision castings. *Modern Casting*, **68**, 5, 57-59.

10. Petch, W. S. (1979). The plaster mold process - its place in the die casting industry. *Proceedings of the 10th International Die Casting Congress*, SDCE, USA, 3 pp.

11. Richards, G. D. (1979). Precision casting. *The British Foundryman*, **72**, 7, 162-167.

## CHAPTER 6

# The Ceramic Mould Casting Processes

## INTRODUCTION

The term ceramic mould encompasses moulds produced by the Shaw process, Unicast process and the lost wax or investment casting process. Each process requires the production of an expendable mould. In the Shaw and Unicast processes a permanent pattern is used to produce a jointed mould, whereas in the investment casting process an expendable wax pattern is used which permits a single-piece, or monolithic, mould to be produced. Similar refractory and binder materials are used in each of the three processes to produce, after firing, strong and inert moulds suitable for casting a wide range of alloys. The castings produced by these processes demonstrate fine detail, smooth surfaces and a high degree of dimensional accuracy [1].

As each of the processes uses similar refractory and binder materials, these are considered in this chapter before a description of the Shaw and Unicast processes is provided. The investment casting process is described separately in Chapter 7.

### Mould Materials

Materials for ceramic moulds must be selected with care if optimum mould properties are to be obtained. In the selection of a suitable material for a particular application the following factors must be considered [2]:

- purity
- refractoriness

- dimensional stability
- thermal expansion
- thermal conductivity
- consistent chemical and physical characteristics
- compatibility with a wide range of alloys
- particle size and distribution
- cost

The mould material must be sufficiently refractory to withstand the pouring temperature of the particular metal being cast without either melting or softening. The melting point value is a good guide to refractoriness, providing that the material is pure. Very small amounts of alkali metal salts or iron oxide can reduce melting points severely. Table 6.1 shows the melting points for a selection of refractories.

Table 6.1 *Refractory Materials and their Melting Points*

| Material | Chemical Notation | Melting Point °C | °F |
|---|---|---|---|
| Magnesia | $MgO$ | 2800 | 5072 |
| Zirconia | $ZrO_2$ | 2677 | 4851 |
| Calcia | $CaO$ | 2600 | 4712 |
| Zircon | $ZrO_2.SiO_2$ | 2420 | 4388 |
| Calcium Zirconite | $CaO.ZrO_2$ | 2345 | 4253 |
| Magnesia Spinel | $MgO.Al_2O_3$ | 2135 | 3875 |
| Alumina | $Al_2O_3$ | 2050 | 3722 |
| Mullite | $3Al_2O_3.2SiO_2$ | 1810 | 3290 |
| Fused Silica | $SiO_2$ | 1710 | 3110 |

The refractory selected should exhibit stability, it should not be susceptible to hydration, or reaction with other materials in the mould or the metal. Thermal stability is also important. Expansion is inevitable. However, providing that it is predictable, i.e. constant and reproducible, it can be taken into account when estimating the contraction allowance for the casting.

Thermal conductivity is a property of the material selected, but it is also a function of particle size and distribution when that material is used to produce a permeable mould. This property influences the rate of heat transfer through the mould and therefore the rate of solidification which may, in turn, influence the metallurgical integrity of the casting.

Cost is an important factor in refractory material selection and is affected by such factors as availability, purity and particle size requirement.

The principal refractories used in the ceramic moulding processes are: alumina, fused silica, alumino-silicates, and zircon.

***Alumina.*** Alumina is produced chiefly from bauxite ($Al_2O_3.2H_2O$). Purification from associated oxides, mainly silica and iron oxide, is by the Bayer process, which consists of dissolving the bauxite in caustic soda and precipitating aluminium hydroxide [$Al(OH)_3$] from the solution by passing carbon dioxide through it. The hydroxide is washed, dried and finally calcined. The process of calcination consists of a high temperature treatment to remove chemically combined water. Because of the use of soda in this process, the final product still contains 0.1-0.2% of $Na_2O$, apart from other residual impurities like $SiO_2$, $Fe_2O_3$, CaO, $TiO_2$ and $Cr_2O_3$ [3]. The presence of $Na_2O$ and $SiO_2$ will reduce refractoriness.

Calcination temperatures over 1000°C (1830° F) ensure that the alumina is in the stable $\alpha$ form. Fused alumina is produced by the electro-thermal fusion of calcined alumina followed by crushing and grading [4].

Tabular alumina is produced by heating the alumina to a temperature near to its melting point. This produces large crystals of $\alpha$ alumina, from which the internal voids have been removed. The material has a high apparent density with low and consistent shrinkage. The purity is also improved by the volatilisation of the residual soda [3]. When used for the investment casting process, the low soda content of the tabular alumina ensures a stable dip coat. However, the higher reactivity of the calcined and fused alumina will produce stronger shells, albeit with higher shrinkage [4].

***Silica.*** Silica exists in several polymorphic forms and significant volume changes are associated with the inversions. Such behaviour would be intolerable in a ceramic mould. However, fused silica, which is produced by melting quartz and solidifying it to form glass before crushing and grading, has an extremely low coefficient of thermal expansion and can therefore be used to produce a dimensionally stable ceramic mould. Fused silica will slowly convert to cristobalite when heated to between 1050 and 1150°C (1920 and 2100°F). This conversion occurs more rapidly above 1200°C (2190°F) and results in an increase in the coefficient of thermal expansion [5].

***Alumino-Silicates.*** Two commonly used alumino-silicates are molochite and mullite.

Molochite is produced by calcining kaolinite (china) clay at a temperature of 1500°C (2730°F) [6]. The clay contains 42 to 43% alumina and 1.5 to 2% alkalis, with the balance being silica. After calcination the material consists of ~56% mullite embedded in an amorphous silica glass. This mineralogy gives a uniform thermal expansion curve up to at least 1400°C (2550°F) and the material softens at around 1600°C (2910°F) [4].

Mullite does not generally occur in nature as a material which can be mined, but is synthesised by heating a mixture of alumina and clay having a composition on calcination equivalent to $3Al_2O_3.2SiO_2$ [5]. The blend may be fired in a kiln to produce the sintered form or melted in an arc furnace to produce the fused form. The latter form is more refractory and more expensive [4].

Less commonly used are the naturally occurring alumino-silicate minerals known as sillimanite, kyanite and andalusite, which are of a similar chemical composition ($Al_2O_3.SiO_2$) but possess different crystal structures and slightly different physical properties. These minerals are approximately 60% by weight alumina and, therefore, would be more refractory than molochite but less so than mullite [4]

***Zircon.*** Zircon ($ZrSiO_4$) occurs naturally as a sand which is available commercially in a chemically purified form. In ground form it is used in primary dip coats and in the natural sand form as a primary stucco for the investment casting process. Its principal advantages are high refractoriness, resistance to wetting by molten metals and a high heat absorption characteristic [7].

The principal characteristics of these materials, together with other materials which also find use in ceramic moulding, are summarised in Table 6.2.

Table 6.2 *Refractory Material Characteristics*

| Material | Specific Gravity | Coefficient of Thermal Expansion/(°C) | Melting Point (°C) | (°F) | Remarks |
|---|---|---|---|---|---|
| Alumina | 3.90 | $9.5 \times 10^{-6}$ | 2050 | 3722 | |
| Fused Silica | 2.21 | $0.5 \times 10^{-6}$ | 1710 | 3110 | |
| Molochite | 2.70 | $4.4 \times 10^{-6}$ | 1600 | 2912 | softening temperature |
| Mullite | 3.14 | $6.0 \times 10^{-6}$ | 1810 | 3290 | decomposes to alumina + liquid silica |
| Sillimanite | 3.23 | - | - | - | |
| Kyanite | 3.60 | - | - | - | |
| Andalusite | 3.15 | - | - | - | |
| Zircon | 4.2 to 4.6 | $4.5 \times 10^{-6}$ | 2420 | 4388 | may dissociate to liquid zirconia + liquid silica at 1600°C |
| Chromite | 4.00 | $7.0 \times 10^{-6}$ | 1400 | 2552 | softening temperature |
| Chamotte | 2.5 to 2.0 | 4 to $6 \times 10^{-6}$ | indeterminate | | |

**Binders**

The most commonly used binders are siliceous and include colloidal silica and prehydrolysed ethyl silicate. The former is used for investment casting, whereas the latter is used for each of the three ceramic moulding processes. Hybrid binders have been developed for the investment casting process and alumina and zircon binders, in colloidal form, have been developed for ceramic shell moulds used in the directional solidification process.

For the investment casting process both air-dried, water-based and ammonia set, alcohol-based binders are used, although water-based binders are preferred for primary coats because of their slower drying characteristics. For secondary coats either system can be used.

***Colloidal Silica Binders.*** These binders are produced by removing sodium ions from an aqueous sodium silicate solution by an ion exchange process [7]. The product consists of a colloidal dispersion of almost spherical silica particles in water. A colloid consists of a suspension of very fine particles in a fluid. In the presence of a liquid, colloidal particles produce a sol. The term sol is used to differentiate them from true solutions on the one hand, and suspensions or slurries on the other [8]. The sol is stabilised by incorporating a small amount of alkali, which creates a negative charge on the silica particles. This causes the particles to repel one another so that they remain separate, preventing agglomeration, see Fig. 6.1.

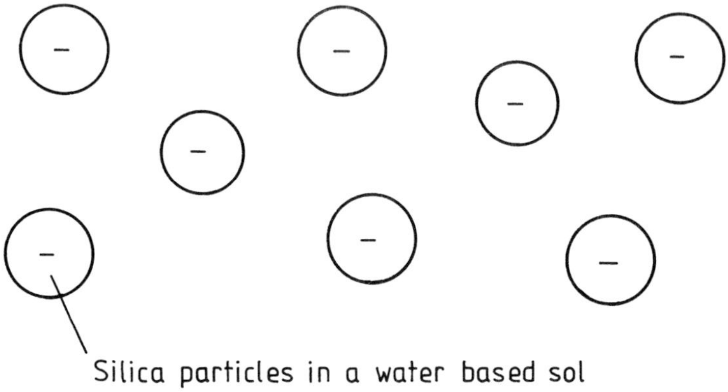


Fig. 6.1 Schematic representation of a stabilised colloidal silica binder

The most popular grades are sodium oxide stabilised and have a silica content of 30% and a particle size within the range 7 to 15 nm (0.3 to 0.6 μin). The binders are used either at this silica content or diluted with water to reduce the silica content to between 18 and 30%. The sols are alkaline with a pH value between 9.5 and 10 and have good storage life. However, they should not be allowed to freeze, otherwise silica is precipitated irreversibly, or allowed to contact iron, as iron oxide will cause local gelation [9].

***Ethyl Silicate Binders.*** Ethyl silicate is produced by the reaction of silicon tetrachloride and ethanol. The product of the reaction is tetraethoxysilane (or tetraethylorthosilicate) with a chemical composition of [$Si(OC_2H_5)_4$]. This material has a theoretical silica content of 28.8 wt %. Commercial ethyl silicates are produced by condensation-polymerisation treatments which eliminate water and increase the silica content. A binder containing 40 wt % silica is generally preferred for ceramic mould production, although lower values will be acceptable if high mould strength is not essential.

Ethyl silicate has no bonding properties unless it is hydrolysed. The reaction with water is usually carried out in ethanol, which is a mutual solvent, using an acid catalyst such as hydrochloric acid. The reaction is fairly rapid and highly exothermic and it forms complex silicic acids that are capable of condensing to form coherent gels having good bonding properties [8, 10].

Prehydrolysed grades of ethyl silicate having reasonable shelf life are available, thus simplifying their use.

## Gelation

Silica gels result from the linking-up of molecules of polysilicic acid, or by aggregation of silica particles of colloidal size [9]. Gelation of silica binders may be effected either by evaporation of the solvent or by chemical means through a change in the pH value.

***Gelation by Solvent Evaporation.*** Colloidal silica binders are gelled, preferably by evaporation of water to deposit silica and develop a green bond. Gelation occurs when the silica particles are brought together. This is most commonly achieved by air drying, which causes the evaporation of water and the concentration of the silica particles. When brought into such close proximity, the force of repulsion is in part negated by that of attraction and the silica gels to provide the green bond for an investment slurry. Ethyl silicates, providing they have been correctly hydrolysed, have exceptionally good drying characteristics and will set firm by evaporation of the alcohol more quickly and efficiently than colloidal silicates.

***Chemically Controlled Gelation.*** The rate of gelation of silica binders is associated with a number of factors, a particularly significant one being pH. The general relationship between the logarithm of gelation time and pH is shown in Fig. 6.2.

Maximum stability, or longest gel time, occurs at a pH value between 1.5 and 2.0. Gelling is most rapid between pH 5 and 6, when the systems set irreversibly. Below pH 1.5 the stability decreases again, while above pH 6 it increases again up to pH 11.

Though slightly acid solutions are the least stable, it is a practical advantage to be able to control gel times by adjusting acidity to within the range pH 4 to 6.

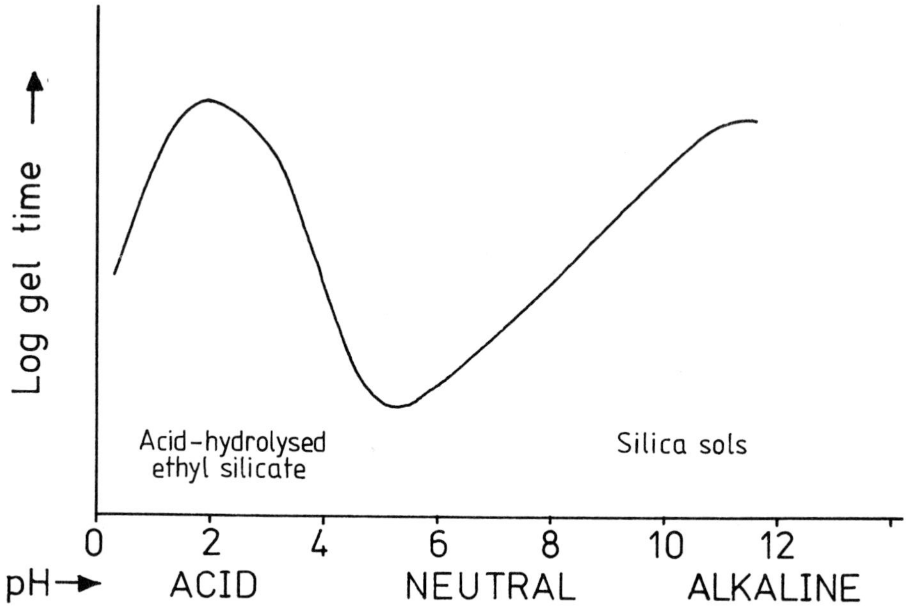


Fig. 6.2 The relationship between the logarithm of gelation time and pH value for silicate binders

Solutions of maximum stability, formed by the acid catalysed hydrolysis of ethyl silicate, lie on the peak of the graph at about pH 1 to 2. The silica sols, on the other hand, are on the alkaline side of neutrality and owe their stability to mutual repulsion of negative charges on the surfaces of the silica particles.

The adjustments necessary to gel the two types of binder solution (acidic or alkaline) are therefore rather different.

With the acid hydrolysed ethyl silicate solutions the pH value is raised to the unstable region where OH ions catalyse the rate of gelation (pH 5 to 6). The alkaline silica sols may be gelled either by reducing their pH to the same region or by adding a solution containing positively charged ions. Neutralising the charge of the particles of such a solution causes extremely rapid gelation, although the actual speed will depend upon the number of positive charges carried by the cations and by the concentration of the solution; adjustment of the pH of the sol does not necessarily cause immediate gelation.

Gelling can be accelerated by using an acid accelerator. For example, a carbon dioxide atmosphere will produce carbonic acid with the water in the binder, neutralising the ionising effect of the stabiliser, which permits the particles to link together causing gelation.

## THE SHAW PROCESS

The Shaw process is a precision casting process capable of the production of accurate castings with excellent surface finish and metallurgical integrity. Moulds are produced using highly refractory aggregates bonded with silica provided by a liquid ethyl silicate binder. A high temperature firing treatment is a feature of the production sequence and this produces an inert mould into which the majority of commercial ferrous and non-ferrous alloys can be cast with confidence.

The process has been used commercially for many years; it was known before the Second World War that silicon esters could be used as refractory aggregate binders [11]. As with most processes there has been a continuous development, in particular with respect to the binder system and the methods of mould production.

### Process Outline

The mould material is prepared by blending refractory powders, containing a high proportion of fine material, with a liquid ethyl silicate binder and a gelling agent. The blended, mobile slurry is poured into the moulding box and around the pattern. Within a short period of time, controlled by the amount of gelling agent, the mould material gels to a rubbery consistency and the pattern can be separated from the mould. On removal of the pattern the mould is torched immediately to remove evolved alcohol. Torching also produces a very fine network of surface and internal cracks, which do not permit metal penetration but may improve permeability and subsequent mould breakdown. After torching, the moulds are fired in a furnace to ensure that there are no combustible materials in the mould and that a strong, rigid, inert, accurate and stable mould is produced.

The advantages claimed for moulds manufactured by the Shaw process include the following [12]:

- good pattern stripping characteristics: the rubbery nature of the gelled mould provides flexibility, which enables the pattern and mould to be separated without damage to the mould when intricate detail, or even straight draws, are required.
- dimensional stability: the excellent reproduction of pattern detail and dimensions is retained by the mould to a great extent after firing and during casting, thus enabling accurate dimensional allowances to be made.
- mould strength: this is sufficient to allow moulds to be cast without the need for moulding boxes.
- collapsibility and resistance to tears: the characteristic internal structure of the mould material improves its breakdown properties and there is less constraint on the casting during contraction.
- resistance to thermal shock: the characteristic internal structure permits expansion of the mould material to occur readily, as a result moulds can safely be poured cold.
- resistance to spalling and washing: the nature of the silica bond prevents the generation of inclusions during mould filling.
- permeability and inertness: as the mould is inert after firing, the only gas to be displaced is that occupying the mould cavity. The characteristic structure provides sufficient permeability to enable the metal to displace the gas readily through the mould.

## Tooling Requirements

As a casting cannot contain more detail or be more accurate than the pattern from which it was made, it follows that particular attention must be paid to pattern construction and quality. The Shaw process utilises permanent pattern equipment and, in principle, a wide range of materials may be used [22]: wood, plaster, graphite, epoxy resin and metal, for example. However, in practice the requirement for high standards of accuracy reduces the choice. Wood patterns are affected by heat and moisture and their dimensions can vary by as much as 3%, which would severely affect accuracy [22]. Plastic patterns and core boxes may be used with the process and may incorporate flexible silicone rubber pieces to produce complex contours and shapes, such as the blade forms in impeller castings [23]. To achieve good dimensional accuracy polished metal patterns are preferred, for which aluminium alloys are particularly suitable. Patterns are normally coated with a release agent which is sparingly applied and well buffed.

## Mould Production

***Mould Material Characteristics.*** The secret of successful mould production and quality castings lies in the mould material mix specification. It is essential to balance the grades of refractory material with the volume of binder and amount of gelling agent in order to produce moulds of consistently high quality. Shaw process slurry preparation is critical [13], too thin and mould cracking will occur on firing, too thick and detail is lost and air bubbles are trapped at the pattern surface. When selecting the grades of refractory material to be used, the principles which apply to sand moulding can be used as a guide. Surface finish will be improved when finer material is used. However, binder requirement will increase as a consequence and permeability will decrease. The strength, measured conventionally by AFS compression and tensile strength, will also be affected by grade selection. Strength will be at a maximum when refractory grades are selected on the basis of using several size gradings, which permit infilling of the voids between larger grains by the smaller [14], this is also the most economic way of using the binder. Mix specification is inevitably a compromise between theoretical considerations and practical requirements, but a reading of standard ceramic texts [15, 16] can be invaluable.

***Mix Specification.*** The choice of refractory and grading will depend upon the factors outlined earlier in this chapter. The ratio of refractory to binder will be dependent upon the density of the refractory and the blend of grades chosen. Practice will therefore vary, but the following procedure has been adopted by the author for the production of castings in die steel [17, 18]. Based on the use of mullite, the following mix specification was adopted:

| | | |
|---|---|---|
| 25% | -8 +16 grade | mullite |
| 25% | -30 +80 grade | mullite |
| 50% | -100 grade | mullite |

The mullite grades were dry blended for one minute prior to adding them to prehydrolysed ethyl silicate on a weight ratio addition of 1 part blended mullite to 0.225 part ethyl silicate. The gelling agent, 10% ammonia in water, was added to the ethyl silicate prior to the blended refractories on a weight ratio of 0.012 part to 1 part blended mullite. Final mixing of the ingredients was completed within one minute, transfer of the mixture to the mould within a further minute, with gelation complete within five minutes of the start of mixing.

***Gelation.*** The principles of gelation were considered earlier in the chapter and for the Shaw process it is the principle of pH control which is usually practised [19]. Ammonia, ammonia salts - acetate, carbonate, hydroxide, or organic ammonia salts - piperidine, mopholemar, triethylamine, may be used for this purpose. The actual time for gelation will depend on the particular application and can be varied by adjusting the amount of gelling agent.

For the preparation of slurries used in the production of large castings [20], magnesium oxide has been advocated as a gelling agent. Addition levels below 2% by weight of the ethyl silicate improved the homogeneity of the slurry and provided a gelation time of between 10 and 15 minutes.

If accurate control of setting time or a very long setting time is required, use may be made of organotin catalysts [21]. The organotin catalysts enable hydrolysis and gelation to proceed readily in the absence of either acid or base. By a special heat treatment process the gelation characteristics can be altered to change the gel time when water is added.

***Mould Production.*** Shaw process moulds may be produced by one of several alternative methods: for small components a ceramic shell is quite acceptable, for larger components a boxless block mould is suitable. If the mould is very large, or if demand for a particular casting is great and it is desirable that mould material costs be reduced, a composite mould can be produced. In this case the facing material is Shaw-based but the backing material might be sodium silicate bonded. For mass production applications the process, in each of these forms, lends itself to mechanisation and rapid output of moulds.

In the production of composite moulds, a procedure for which is shown in Fig. 6.3, additional pattern requirements in the form of a pre-form pattern may be necessary. These may be produced in wood, as their purpose is only to produce an oversize cavity in the backing mould. This mould is then positioned over the precision pattern and the gap filled by introducing the mobile Shaw slurry. Permanent pre-form patterns are not essential, the required effect can be obtained by covering the precision pattern with a thickness of material such as oil-sand, $CO_2$/silicate sand, felt, etc [23-26], which will permit the oversize backing mould cavity to be produced. To ensure that separation of the backing and facing layers does not become a problem, the mould materials selected for the backing and facing layers should have similar thermal expansion characteristics. In a recent development [23], a technique based on the slip casting of a refractory aggregate into plaster moulds has been used to produce contoured backing shell moulds for composite Shaw mould production. Permanent metal backing moulds have also been proposed [27].

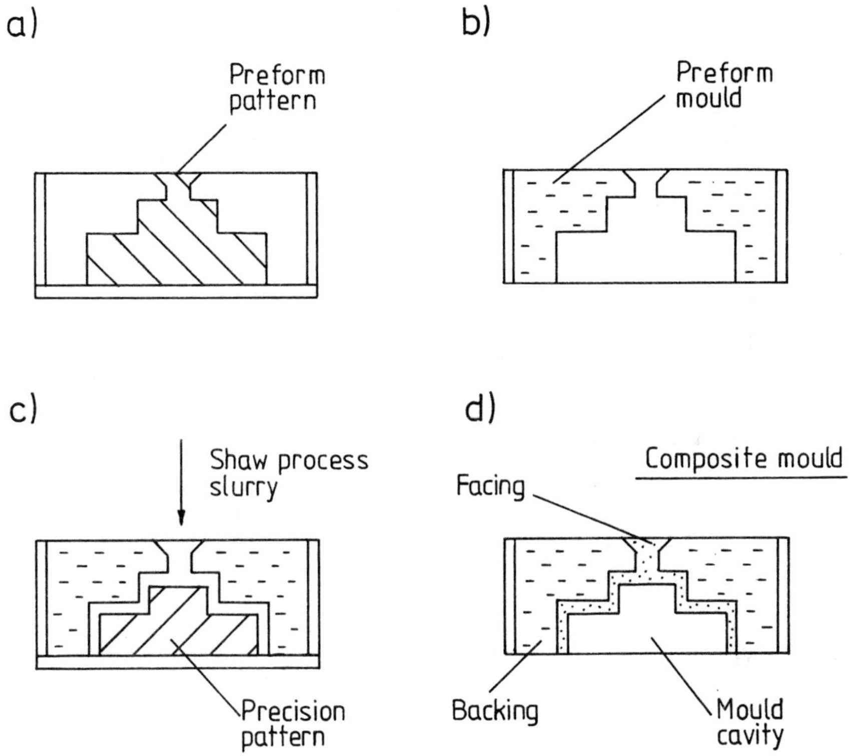


Fig. 6.3 Stages in the production of a composite Shaw process mould

Figure 6.4 shows a selection of block and composite moulds produced by the Shaw process.

***Mould Processing.*** The separation of the pattern from the mould is a source of dimensional inaccuracy in most moulding processes. It is an aspect which requires particular care and attention in a precision moulding process. To provide consistent pattern stripping a mechanised action is recommended [18]. When the mould has been separated from the pattern it must pass through several stages which can exert a significant influence on the quality of the finished mould. The craze cracking effect in Shaw process mould materials is only produced when ignition follows immediately after pattern stripping. When the mould material is allowed to air-dry for a substantial period, the craze cracking does not develop. It is interesting to note that the strengths developed by firing are considerably higher in the air-dried material - but at the expense of permeability, thermal shock resistance and breakdown characteristics [28].

Fig. 6.4 The production of moulds, including composite moulds, by the Shaw process

Coarse cracking may be a problem at internal corners in a mould cavity, particularly if pools of alcohol are allowed to form. Sharp corners are unsuitable in most casting designs and a minimum radius of 3 mm (0.125 in) is recommended where practicable. The evolved alcohol usually ignites readily but ignition can be encouraged by the use of a gas-air torch, in either case the moulds should stand on a grid to provide free access of air during ignition.

After torching, moulds are fired for two reasons:

(a) To burn off any residual organic material and remove any moisture in order to produce an inert mould.

(b) To improve mould strength through a sintering effect.

A considerable range of baking or firing temperatures, from 300 to 1300°C (570 to 2370°F), is suggested in the literature. Whereas the first objective should be met by "firing" at 500°C (930°F), it is unlikely that any substantial improvement in strength will be obtained by a sintering temperature below 1000°C (1832°F). Slightly oxidising conditions are recommended [29] when firing moulds to encourage the burning of residual volatiles. Composite moulds present their own difficulties, with radiant heating in tunnel furnaces being preferred to fire the facing material without excessively heating the backing material [30].

**Mould Preparation for Casting**

Moulds may subsequently be cast whilst still hot, or allowed to cool, depending on section thickness and metallurgical requirements. Mould part location in boxless moulds is obtained either by double-ended conical pins, which sit in prints moulded in each mould half, or by stepped joints. Mould joints are sealed externally with a luting compound, joint face seals reduce accuracy across the joint, and moulds may be weighted or clamped using metal bands before casting.

***Unicast Process Variation.*** The Unicast process varies from the Shaw process in the method of mould stabilisation employed. Whereas in the Shaw process stabilisation is achieved by ignition of the mould immediately after stripping, in the Unicast process the green mould is immersed in a bath of stabilising liquid before torching and subsequent firing [31]. Various stabilising media can be used and include: ethyl alcohol, acetone, benzene, kerosene, and even boiling water [1]. The stabilisation procedure is claimed to eliminate mould warp and distortion. Special equipment for the mass production of moulds by the process is supplied by the Unicast Development Corporation and includes equipment for slurry preparation, pattern stripping, mould stabilisation, and mould firing [32].

It has been claimed that the surface quality of moulds, and consequently the surface finish of castings, can be improved by dipping the moulds in a solution of calcium chloride before torching [20].

**Gating and Feeding**

The usual principles of gating and feeding practice apply. However, top gating, with metal poured through the head, is possible with the process. The insulating nature of the mould material, especially when preheated, prolongs the solidification time of the casting. This may require the use of large feeder heads to ensure that they remain liquid longer than the casting.

**Metallurgical Considerations**

The nature of Shaw process mould material results in relatively slow cooling of the metal in comparison with other casting processes. This can cause a coarse grain structure which may be unacceptable in certain castings' applications: it can be countered by grain refining additions. The slow cooling is, in part, responsible for decarburisation and surface oxidation in steel castings [33]. In steels containing chromium the surface may be further affected by pitting [34].

Decarburisation has been shown to be a function of the time that the casting remains at an elevated temperature, that is above 875°C [33]. In practice, the severity of the problem can be reduced by one of several methods, which include: hexamethylene tetramine placed on the exposed surface of the metal when the mould is enclosed within a container [35]; a reducing gas atmosphere [36]; the incorporation of graphite powder in the mould facing material; or, most effectively, by a vacuum treatment to remove air from the mould followed by back filling in an inert gas [33]. The occurrence of pitting has been shown to be associated with decarburisation [34]. The two phenomena are caused by carbon monoxide formation. Carbon in the metal reacts with oxygen from the mould atmosphere and the silica which provides the mould bond. This results in decarburisation and the carbon monoxide bubbles deform the casting surface. This implies that it may not be possible to eliminate decarburisation entirely because oxygen will always be present in the silica which provides the bond.

It has been suggested that casting surface characteristics can be altered by passing a gas through the mould material as a steel casting is cooled. By using ammonia [37], a nitrided surface can be developed and a carburised surface produced by using a carbon rich atmosphere [29].

Heavily oxidised surfaces reduce both dimensional accuracy and the quality of surface finish directly and through the requirement for abrasive shot blasting. The use of inert gas atmospheres to minimise these effects can be worthwhile.

**Casting Finishing**

Finishing is by conventional means.

## Reclamation

The reclamation of mould materials, especially from composite moulds, is not recommended. To ensure consistency, new materials should always be used for the facing material.

## Health and Safety Aspects

Inhalation of the fine dust associated with the mix formulation and knock-out of the moulds should be avoided. The binder fumes are hazardous to health when inhaled and the presence of an alcohol carrier presents potential explosive and/or fire risks. Protective face masks and well ventilated working areas are recommended.

## Process Capabilities

***Alloy Range.*** A wide range of ferrous and non-ferrous alloys can be cast using these processes. The range of ferrous alloys cast includes: carbon and low alloy steels, stainless steels and tool steels. Whilst ferrous production is predominant, aluminium-base alloys and copper-base alloys, including beryllium-copper, are cast by the process.

***Casting Size and Weight Range.*** Whilst small castings can be produced by the process, it is for those castings considered outside the range of investment casting that the process is most noted. Castings weighing several tonnes, typically dies and tools, are routinely produced by the process.

***Production Volumes and Lead Times.*** The process is most suited to small quantity production. Mould production is quite slow, although it is claimed that, with mechanisation, an output of fifteen moulds/hour could be achieved from a single set of pattern equipment [38]. Although a precision pattern is required, lead times are shorter than for investment casting or pressure diecasting and the tooling costs are lower.

***Thin Section Capability.*** Whilst most production consists of heavier sectioned castings, thin sections can be cast. With special attention to process procedure, sections of 1.5 mm (0.06 in) could be cast.

***Surface Finish.*** The surface finish of a casting produced by a ceramic moulding process is influenced by several factors, which include: refractory grading, metal and mould temperature, and metal mould reactions. The finer the grade of refractory filler used in the refractory aggregate, the better should be the surface finish. Preheated moulds and high casting temperature are likely to reduce the quality of surface finish [39]. Generally, castings can be produced with surface roughness values better than 3.2 μm (125 μin).

***Dimensional Accuracy.*** Attainable accuracy in relation to size of dimension is shown in Table 6.3. The figures refer to steel castings and do not include an allowance of between ± 0.25 to 0.50 mm (± 0.01 to 0.02 in) for dimensions measured across the joint.

Table 6.3 *Attainable Dimensional Accuracy in Castings Produced by the Ceramic Moulding Process*

| Dimension Size | | Tolerance | |
|---|---|---|---|
| mm | in | mm | in |
| up to 25 | up to 1 | ± 0.08 | ± 0.003 |
| 25 to 75 | up to 3 | ± 0.13 | ± 0.005 |
| 75 to 100 | 3 to 8 | ± 0.38 | ± 0.015 |
| 200 to 375 | 8 to 15 | ± 0.76 | ± 0.030 |
| over 375 | over 15 | ± 1.14 | ± 0.045 |

***Typical Castings.*** Typical of the output in steel are cast to shape/size dies and tools for extrusion, forging, die casting, plastics and glass production [1]. Recent interest has extended to cast dies for advanced composite component production [40]. Other applications include: aircraft and aerospace components; marine fittings and propellers; complex parts for gearing, precision machinery and pumps; impellers and valves; and components for the leisure industry [41].

Figure 6.5 shows a selection of castings produced by the Shaw process.

Fig. 6.5 Photograph showing a selection of castings produced by the Shaw process

## References


1. Anon. (1988). Ceramic moulding. In: *Metals Handbook, Vol. 15 Casting*, 248-252. Published by ASM.

2. Bidwell, T. (1969). *Investment Casting*. Published by the Machinery Publishing Co. Ltd., Brighton.

3. Waye, B. E. (1967). *Introduction to Technical Ceramics*. Published by Maclaren, London.

4. The BICTA Materials and Processes Committee (1974). *Investment Casting: Mould Refractories and Suppliers*. Published by BICTA, 13 pp.

5. Singer, F. and Singer, S. S. (1963). *Industrial Ceramics*. Published by Chapman and Hall, London.

6. *Molochite*. Product brochure published by ECC International Ltd.

7. Horton, R. A. (1988). Investment casting. In: *Metals Handbook, Vol. 15 Casting*, 253-269. Published by ASM.

8. Mills, D. (1989). Understanding silica binders. *Foundry Trade Journal,* **163**, 3389, 288-290.

9. *Binders and Applications in the Precision Foundry Industry.* Product brochure published by Monsanto Europe SA.

10. Emblem, H. G. (1979). Preparation and properties of ethyl silicate. *Proceedings of the 15th Annual BICTA Conference,* Paper No. 1, 10 pp.

11. Shaw, C. (1946). The application of ethyl silicate to foundry practice. *Foundry Trade Journal,* **78**, 1534, 31-33.

12. Lubalin, I. and Christensen, R. J. (1960). Shaw process patterns, principles and production. *Modern Casting,* **38**, 4, 83-90.

13. Paterson, S. (1967). Precision cast hot sizing dies. *SAE Transactions,* **75**, Paper 660645, 263-271.

14. Lawrence, W. G. (1960). Precision casting mold materials. *Modern Casting,* **38**, 2, 109-118.

15. Kingery, W. D. (1976). *Introduction to Ceramics.* Published by Wiley, London.

16. Budworth, D. W. (1970). *An Introduction to Ceramic Science.* Published by Pergamon Press, Oxford.

17. Clegg, A. J. (1980). The Shaw process - a review. *Foundry Trade Journal,* **149**, 3197, 429-438.

18. Clegg, A. J. and Das, A. A. (1987). The production of precision cast H13 die steel tooling by the Shaw process. *The British Foundryman,* **80**, 3, 137-144.

19. Lawrence, W. G. (1960). Precision casting mold materials. *Modern Casting,* **38**, 2, 109-118.

20. He Xiang Ping (1989). Precision cast dies produced by a ceramic mould process - a review. *Cast Metals,* **2**, 1, 16-19.

21. Jones, K., Biddle, K. D., Das, A. K. and Emblem, H. G. (1979). Organotin catalysts for organic silicate binders. *Tin and its Uses,* 199, 10-11.

22. Butler, M. J. (1970). Cast moulds: ferrous and non-ferrous materials. *British Plastics,* **43**, 6, 144-148.

23. Ball, W. R. (1990). Private communication.

24. Winter, D. B. (1964). Shaw process moulds at Innsworth Metals. *Foundry Trade Journal,* **116**, 2474, 427-434.

25. Dittrich, W. (1972). Precision casting processes with ceramic moulds: a review of the present situation and possible future developments. *Giesserei,* **59**, 8, 237-249.

26. Dunlop, A. (1960). Mold structure and its production. *US Patent* 2,931,081.

27. Lubalin, I. J. (1966). Molds and a method of preparing same. *US Patent* 3,242,539.

28. Shaw, N. (1957). Method of making molds. *US Patent* 2,795,022.

29. Olsen, C. F. (1969). The use of precision casting as a prototyping tool. *ASTME Technical Paper,* CM69-235.

30. Prasad, J. S. and Watmough, T. (1969). Precision-cast superalloy dies for isothermal forging of titanium alloys. *Trans AFS,* **77**, 289-296.

31. Greenwood, R. E. (1965). Method of making ceramic molds. *US Patent* 3,172,176.

32. Greenwood, R. E. (1969). Ceramic molding in the pattern shop. *Trans AFS,* **77**, 377-379.

33. Doremus, G. E. and Loper, C. R. (1970). A study of decarburization accompanying the casting of steel in ceramic molds. *Ibid,* **78**, 338-342.

34. Yardimici, G., Fischer, F. and Klein, F. (1976). Surface layer decarburization and pitting in steel investment castings. *Proceedings of the 4th World Conference on Investment Casting*, Paper No. 11, 17 pp.

35. Shaw, C. (1960). Prevention of scale formation on castings. *US Patent* 2,935,722.

36. Lubalin, I. (1965). Cast forging dies. *Paper presented to ASTME,* Hartford, Connecticut, USA, 19 January 1966.

37. Pikman, A. A., Lyakhovich, L. S., Polyakov, B. Z. and Surkov, V. V. (1976). Nitriding of ceramic-mould castings. *Russian Castings Production,* 253-254.

38. Greenwood, R. E. (1976). High-volume production of precision castings by ceramic molding (Unicast process). *Trans AFS,* **84**, 417-420.

39. Forminykh, I. P., Chen, B. S., Yakshin, N. M., Kavitskii, I. M., Cherkasov, V. E. and Bobrovskikh I. E. (1976). Quality improvement in cast dies. *Russian Castings Production,* 467.

40. Stein, D. and Peters, D. T. (1989). Cast-to-size aluminium-bronze tooling for fabrication of advanced fibrous composite aircraft components. *Int. J. of Materials and Product Technology,* **4**, 4, 334-350.

41. Anon. (1980). Shaw process technology at Precision Products (Cumberland) Ltd. *Metallurgia,* **47**, 6, 316-322.

CHAPTER 7

# The Investment Casting Process

## INTRODUCTION

The Investment Casting Process, also known as the lost wax process, is probably the process most commonly associated with the term precision casting. It is unique in the fact that its applications span the millenia, featuring in ancient civilisations and space-age technology. It is possible that the first investment casting was produced in Thailand around 4500 BC [1]. Archeological evidence exists to show that the process was used in Mesopotamia during the period 4000 to 2000 BC; in China, India, Greece and Northern Europe during the period 2000 BC to Anno Domini; in Italy, Central and South America, West Africa and Western Europe in the period from Anno Domini to 1000 AD [1]. Whilst noted for its use in the production of statuary and jewellery, the process was also used in the production of weaponry, tools and utensils.

The development of the process which has evolved into today's manufacturing technology is credited to Austenal Laboratories Inc in the USA [2]. In 1932 the organisation developed the use of a hydrolysed ethyl silicate binder to produce moulds into which their high melting point dental alloys, based on cobalt and chromium, could be cast. The process and alloys were extended to the production of aircraft turbo-supercharger parts for GE in 1940. It was the suitability of the process for the manufacture of heat resisting alloy components for the aero-engine industry which spurred its development. By the late 1940's/ early 1950's the technology for the ceramic shell process was developed and is now used almost universally by the investment casting industry [3].

**Market Trends**

In 1945 the sales value of investment castings reached $35 million in the USA as a result of the war effort, but after the war demand slumped and in 1948 the sales value in the USA was only $5 million [4]. The post-war growth of the aero-engine market enabled sales to increase so that by 1958 they were $70 million, in 1979 $1 billion, and by 1987 exceeded $2 billion. The market in the USA is dominated by the aerospace industry, which represents around 65% of the market. According to Stratton [5], of the 6000 foundries in business in the USA in 1988, 350 were investment casters. Although this represents only 6% by number, these foundries accounted for 17% of the $15 billion US casting market. The US industry has grown since 1946, and is expected to continue growing, at the rate of 6 to 9%/annum.

Growth in the UK has been at a similar rate. In 1958 the sales value of the UK industry was estimated at £5 million. By 1968 the value was £11 million, in 1978 £49 million, and by 1988 approached £250 million [4, 6]. In 1988 there were 55 investment casters in the UK, in comparison with 80 in mainland Europe and 65 in Japan [6]. In terms of world market share the USA leads with a share of 60%, the UK and Europe each has 12%, and Japan 8% [7]. This excludes Eastern Bloc countries, China, and the strongly developing activities of Taiwan and Korea.

A forecast published in 1990 [6] suggests that the £2 billion world market for investment castings in 1989 will have increased to £2.75 billion by 1995. The distribution of this market by alloy type is shown in Table 7.1.

Table 7.1 *The World Market for Investment Castings [6]*

| Alloy Groups | Market Share (%) | |
|---|---|---|
| | 1989 | 1995 |
| Air Melt and Vacuum Melt Superalloys | 46 | 36 |
| Stainless Steels | 31 | 37 |
| Carbon Steels | 6 | 14 |
| Titanium | - | 4 |
| Aluminium | - | 6 |
| Others | 17 | 3 |

**Process Outline**

The starting point for the process is the production of an expendable (usually wax) pattern. This pattern is a precise component produced from a precision-engineered die. Complex features may be obtained in the pattern by the use of multi-part dies or by the use of soluble wax cores or permanent ceramic cores. The wax patterns are assembled with the gating and feeding system and cleaned prior to their investment with the ceramic coating. This ceramic coating is built up through successive stages of dipping and stuccoing. The initial (primary) dip coat, containing a fine refractory, is allowed to gel before the assembly is dipped in a secondary dip tank and stuccoed with coarser, dry particles in a raining cabinet or fluidised bed. Following gelation of the binder this procedure is repeated until the required shell thickness is obtained. On completion of the gelation stage the expendable wax pattern is removed, preferably in a steam autoclave, and the ceramic shell fired to around 1000°C (1832°F) prior to casting.

**Tooling Requirements**

Precision tooling for the production of wax patterns is expensive, but its cost is low in comparison with tooling for pressure diecasting. Waxes are processed at low temperatures and pressures and lack abrasiveness. Consequently, the demands placed on the tool material are not great. The majority of tools are produced by machining from wrought material. Aluminium alloys are generally preferred, they are easy to machine, have good thermal conductivity, good corrosion resistance, and are relatively light in weight and easy to handle. If improved wear resistance is required in certain parts of the die, for example where sliding parts are used or accurate location is required, brass or steel inserts may be used. For the production of plastic patterns, such as polystyrene, tool steels are preferred, with P20 a popular choice.

A variety of techniques, other than machining, may be employed in the production of tooling and include: cast tooling, hobbing, spraying, and electroforming [8].

**Pattern Waxes**

Waxes are organic compounds of carbon, hydrogen and oxygen. They are esters of fatty acids and monohydroxylic alcohols. Esters are produced when alcohols react with acids, including organic acids, and lose a molecule of water in the process. Examples of natural waxes include Carnauba wax, which coats the leaves of the Carnauba palm, native to Brazil, and is largely myricyl cerotate, $C_{25}H_{51}CO_2C_{30}H_{61}$; and beeswax, which is largely myricyl palmitate, $C_{15}H_{31}CO_2C_{30}H_{61}$. Waxes are generally harder, more brittle and less greasy to touch than the fats and oils which, together with waxes, are termed lipids. Lipids are all insoluble in water and highly soluble in one or more organic solvents including ether, chloroform, benzene and acetone. Waxes are thermoplastic in behaviour, which means they can readily be melted and cast into shape.

Historically, pattern waxes were based on beeswax, although these were often modified with natural resins (which would harden the wax) or clarified butter (which would soften the wax) [9]. However, modern investment casting waxes are complex compounds containing numerous components which include: natural hydrocarbon waxes, natural ester waxes, synthetic waxes, natural and synthetic resins, organic filler materials, and water [10]. By varying the combinations and proportions of these materials, properties such as melting point, viscosity, expansion/contraction, setting rate, hardness, etc, can be manipulated to suit specific requirements.

## Wax Classification

As indicated previously, a variety of materials are used under the heading of waxes. These may be classified within three broad categories [11]:

Natural Ester Waxes
Natural Hydrocarbon Waxes
Synthetic Waxes

## Natural Ester Waxes

Natural ester waxes are used in small amounts, generally $<10\%$, to increase the hardness of the blend and accelerate the setting process when the wax cools. Being natural products they may have inconsistent properties and they have a high ash content. The price of these natural waxes is often very high. Carnauba has a fairly high melting point of 83°C (182°F), a low coefficient of thermal expansion and is hard, brittle and non-tacky [8]. Candelilla is another vegetable wax, obtained from the stems of a weed which grows prolifically in Mexico, which has a relatively low coefficient of thermal expansion. It is moderately hard but slightly tacky. It exhibits less solidification shrinkage than the hydrocarbon waxes and has been used for hardening paraffin waxes and raising their softening point. Beeswax has a melting point of ~64°C (~147°F) and because of its relative softness and high price is rarely used in modern pattern waxes.

### Natural Hydrocarbon Waxes

The most commonly used hydrocarbon waxes are those derived from petroleum crude oils. These can be sub-divided into: paraffin waxes, intermediate waxes and micro-crystalline waxes.

***Paraffin Waxes.*** These waxes are usually produced from relatively low boiling point lubricating oil by a vacuum distillation process [11]. They are available in closely controlled grades with melting points varying by ~3°C (~5°F) increments. However, those with melting points in the range from 52 to 68°C (125 to 155°F) are most commonly used [8]. The material is fairly translucent when highly refined and exhibits a coarse crystal structure when fractured. The low cost of these waxes, combined with their ready availability, convenient choice of grades, high lubricity and low melt viscosity, accounts for their wide usage. However, applications are limited by their brittleness and high shrinkage.

***Intermediate Waxes.*** These waxes can be referred to as semi micro-crystalline materials. They are obtained from the highest boiling point lubricating oil and are also produced by vacuum distillation. These waxes are generally opaque, hard, have a finer crystalline structure than the paraffin waxes, and are less brittle [11].

***Micro-Crystalline Waxes.*** These waxes are obtained from the residues of lubricating oil, the distillation being carried out under high vacuum. They have a very much higher molecular weight than paraffin waxes. These waxes are opaque and have a fine crystalline structure [11].

### Synthetic Waxes

These materials are man-made derivatives of either natural ester or natural hydrocarbon waxes. They include: Fishertropsch waxes, polyethylene waxes, polyethylene glycol waxes, derivatives of Montan waxes, oxidised waxes, emulsified waxes, and chlorinated waxes [11].

Waxes are easily blended to suit different requirements. Their low melting points and low melt viscosities make them easy to compound, inject, assemble into clusters, and melt out without cracking the thin ceramic shell moulds. These properties allow waxes to be injected at low temperatures and pressures and this, combined with their lack of abrasiveness, leads to lower tooling costs [8].

However, although plain waxes possess these useful properties, there are two important areas in which they are deficient:

1. Strength and rigidity, especially where very fragile patterns need to be made.

2. Dimensional control, especially with regard to surface cavitation resulting from solidification shrinkage during and after pattern injection.

Improvements in these areas can be made with non-wax additives.

**Additives**

Improvements in the strength and toughness of waxes are usually obtained by the addition of polymers. Polyethylene is the most widely used addition because it is compatible with a wide range of waxes and relatively inexpensive. Other materials which can be used include: ethylene vinyl acetate, ethylene vinyl acrylate, nylon, and ethyl cellulose [8]. The amount of these materials that can be added is limited because they are viscous at the temperature at which the waxes are used. However, significant improvements in the pattern handling characteristics can be obtained by the small additions that can be tolerated.

Surface cavitation, or sinking, is a manifestation of solidification shrinkage. The addition of polymers will reduce the amount of shrinkage in proportion to the amount added. However, as the polymer addition is relatively small the effect is also small. To achieve a more pronounced effect, resins or fillers may be added to the wax. The former are widely used, whereas the latter are used selectively. Waxes are therefore described as being either filled or unfilled [8].

***Resins.*** A variety of resins are available and include synthetic resins and those derived from natural sources, such as pine trees, crude oil and coal tar. The choice depends upon their compatibility with the waxes and polymers used, their softening point and viscosity versus temperature relationship. An essential requirement is that they exhibit a small volumetric change, in comparison with the wax, when heated or cooled through their melting/solidification range. Consequently, their addition reduces shrinkage in direct proportion to the amount used.

***Fillers.*** Whereas resins are liquid at the wax processing temperature, fillers are not. As the filler remains a solid it does not contribute to the solidification shrinkage of the wax blend. Hence, the total shrinkage of the wax mixture is reduced in direct proportion to the amount of filler added.

An ideal filler should be organic, leave no ash, have a relatively high melting point, be non-reactive towards both the wax mixture and the shell binder, have a fine particle size, be similar in specific gravity and have a similar thermal conductivity to the wax mixture. The most commonly used filler in the UK is probably crosslinked polystyrene, with isopthalic acid and water being the only others used with any regularity. Water can be used in the form of emulsified waxes containing between 7 and 12% water. At the temperatures used in wax processing the water remains a liquid and, therefore, does not contribute any shrinkage. In the USA crosslinked polystyrene is rarely used. The preferred fillers are isopthalic or terepthalic acid, although bisphenol and poly alpha methyl styrene are also used [9].

## Wax Selection

The selection of a suitable wax is determined by the particular application, but the factors which should be considered at each of the processing stages are summarised in Table 7.2 [8].

Table 7.2 *Wax Selection Criteria*

| | |
|---|---|
| Injection | Softening point, freezing range, rheological properties, ability to duplicate detail, surface, and set-up time. |
| Removal, Handling and Assembly | Lubricity, strength, hardness rigidity, impact resistance, stability, and weldability. |
| Dimensional Control | Thermal expansion/contraction, solidification shrinkage, cavitation tendency, distortion, and stability. |
| Mould Making | Strength, wettability, resistance to binders and solvents. |
| Mould Dewaxing and Burnout | Softening point, viscosity, thermal expansion, thermal diffusivity; and ash content. |
| Miscellaneous | Cost; availability; ease of recycling; toxicity; and environmental factors. |

## Water Soluble Waxes

Soluble core waxes are comprised of a minimum of three ingredients: a binder, filler and effervescing agent. The fillers and effervescing agents are inorganic, so extreme care must be taken to ensure that contamination of the pattern wax does not occur. The binder is generally a polyethylene glycol, which is available as a blend of molecular weights used to obtain the desired flow and softening point. These materials are slow to dissolve in aqueous solutions and the solubility decreases with increase in molecular weight. The filler is generally a powdered material which is present to improve strength and decrease shrinkage. Silica, mica and sodium chloride are commonly used as fillers. Effervescing agents are added to the formulation to aid the solubility of the polyethylene glycol. Carbonates are commonly used and react with the acid in the leaching solution to form gaseous carbon dioxide, which increases the rate at which the wax dissolves [12].

Leaching solutions are weak concentrations of acids, typically hydrochloric acid, although sulphuric, nitric and acetic acids have all been used. Due to the chemistry of the leaching operations, the acid becomes progressively weaker and further additions are necessary to regenerate the bath. However, too high a concentration can cause the surface of wax patterns to become tacky [13].

## Alternative Pattern Materials

Waxes are not used exclusively as pattern materials. Plastics, such as polystyrene, are used because their strength reduces damage in thin section patterns. However, although they can be readily produced and appear attractive, they have a tendency to cause mould cracking at the pattern removal stage. The problems of this dense polystyrene can be overcome by the use of expanded polystyrene. The latter material has been used for gating systems and is used in the Replicast CS process, but is not a realistic alternative to wax. Other alternatives to wax include: urea-base materials, which are readily injection moulded and soluble in water; mixtures of low melting point salts, which are also water soluble; paradichlorobenzene; and napthalene [8].

## Melting Waxes

Care should be taken when melting waxes prior to conditioning or injection [14]. Waxes should never be overheated as this may oxidise some of the constituents, embrittle the wax, or affect its injection characteristics and dimensions. Wax should be kept at a temperature between 80 and 90°C (175 and 195°F) during melting.

Waxes are best melted in an oil-jacketed container that is temperature controlled by a reliable thermostat. In addition, a slow speed stirrer should always be used while the wax is in a molten state. This is essential when melting filled waxes. These recommendations will help maintain uniform temperatures and prevent localised overheating and damage to the wax. To minimise filler separation, a cylindrical tank (no corners) should be used, with the lower blade of the stirrer as close as possible to the base of the tank.

**Wax Conditioning**

Waxes are complex mixtures which must be homogeneous before use [14]. To achieve this requirement and optimise viscosity, flow and surface finish, the wax should be held at the injection temperature for several hours before use. Care should be taken to ensure that the wax in the melting tank is at the same temperature as that in the injection machine reservoir.

**Wax Injection Equipment**

Wax injection machines are designed to take preconditioned wax and force it into a die to make a pattern. There are three basic types of injectors: liquid, paste and solid. As the temperature reduces, the viscosity of the wax increases, and more pressure is required to force the wax into the die.

The liquid machine consists of a wax reservoir (which usually contains a wax mixing unit), injection cylinder, injection nozzle, and valve assembly [15]. The wax flows from the reservoir to the injection cylinder, where it is pressurised and forced out of the nozzle into the die. The liquid wax injector is by far the most popular machine used today, despite some of the inherent problems associated with using a liquid wax. The first of these is shrinkage, which increases with an increase in liquid wax temperature. However, this can be reduced by the incorporation of fillers. The second problem to overcome is the entrapment of air caused by turbulence in a low viscosity, hot liquid wax. This can be overcome by exercising flow control using electronically controlled flow devices. The principal variables to be controlled during wax injection are: temperature (wax and nozzle), pressure and flow. To this may be added the need to control dwell time and die clamping pressure. An example of a fully automatic wax injection machine is shown in Fig. 7.1.

Fig. 7.1 Photograph of a fully automatic wax injection machine with multiple impression tooling

## Quality Control Tests for Pattern Waxes [16]

***Softening Point by Ring and Ball.*** The ring and ball softening point is the temperature at which a wax sample held in a horizontal ring is forced downward a distance of 25 mm (1 in) under the weight of a steel ball as the sample is heated at a set rate in a water or silicone oil bath.

***Needle Penetration Hardness.*** The penetration is the depth in mm (in) that a standard needle penetrates vertically into a sample under set conditions of load, time and sample temperature.

***Total Ash Content.*** Total ash is the percentage of residue left after complete combustion of a sample.

***Specific Gravity.*** This is the weight per unit volume.

***Kinematic Viscosity.*** This is the measure of the time for a fixed volume of liquid to flow by gravity through a capillary.

***Dynamic Viscosity.*** This is the numerical product of kinematic viscosity and the density of the liquid, both at the same temperature.

**Pattern Assembly and Preparation**

A pattern should preferably be produced as a single entity. However, where size or complexity preclude this, it will be necessary to assemble the pattern from segments. Small patterns are usually produced as a single piece, often complete with gates, but they, like the larger patterns, will require assembly in order to produce the final cluster. Extensive use is made of preformed wax gating/feeding systems and of preformed ceramic pouring cups. The assembly of the clusters is usually a manual operation and the wax components are readily assembled by wax welding using a heated spatula. It is essential to ensure that all wax joints are fully sealed. The ingress of the ceramic coating into incorrectly sealed wax joints is a potential source of inclusions in the casting.

Following assembly the clusters are cleaned to remove mould release agent (often silicone-based and therefore not wet by water), loose pieces of wax or dirt. The clusters are washed in water containing a wetting agent or a solvent solution that does not attack the wax. In some cases a solvent mixture capable of attacking the surface of the wax may be used, as the controlled etching effect may improve slurry adhesion without adversely affecting the surface quality of the casting [8]. After cleaning or etching a clean water rinse is required to remove the cleaning/etching agents.

**Mould Manufacture**

The moulds used for investment casting can be produced either as block moulds or ceramic shells. The process was originally established using block moulds, but this procedure has largely been superseded by the ceramic shell process.

***Block Moulds.*** The wax pattern is dipped using a primary coat which contains fine refractory particles. Whilst still wet, coarse particles are applied as a stucco and the assembly is allowed to dry. The coarse particles act to reduce drying stresses and promote the keying of the backing material. The dried assembly is fastened to a mould base plate to which a mould flask is attached. The backing investment is then poured into the flask and vibrated to ensure close contact with the coated pattern and achieve consolidation. After gelation the mould is dewaxed, fired and cast.

***Ceramic Shell.*** In this process the mould is produced by a series of successive dipping and stuccoing operations to build up a shell of a thickness sufficient to provide the strength required to process the mould successfully. The primary dip is usually different to the secondary or backing dip, and the grade of the stucco particles used may be coarser as the shell thickness increases. Although manual manipulation of the clusters is practised during dipping and stuccoing, increasing use is being made of dedicated manipulators or robots. As the size and weight of clusters increases their use becomes essential. A further advantage is their repeatability of action, which improves the consistency of shell build-up.

## Slurry Formulation, Preparation and Control

***Slurry Formulation.*** The slurries used for the primary and secondary dip coats contain a fine refractory powder and a liquid silicate binder as the principal ingredients. However, other ingredients may be present and include: wetting agents, antifoam compounds and strength enhancers [8]. The former is usually confined to colloidal silica primary coats and, in amounts between 0.03 and 0.30 wt % of the liquid, promotes wetting of the pattern by the primary coat. The antifoam compounds are commonly used in association with the wetting agent to suppress foam formation and promote air bubble escape, especially from the pattern surface. Additions of 0.002 to 0.10 wt % of the liquid are sufficient. The strength enhancers may be used both to enhance the coating ability of the slurry and improve the green strength and resilience of the dried coating. A variety of materials are available to meet these requirements [8].

***Slurry Preparation.*** The slurries are prepared by adding the refractory powder to the binder liquid. Agitation is necessary to break up agglomerates and thoroughly wet and disperse the powder. A prolonged period of mixing is required to promote de-aeration and particle wetting. Once homogenisation has been achieved stirring must be continued, albeit at a reduced level, to ensure the powder remains in suspension. Although propeller mixers are used, there is a preference for the use of rotating tanks which incorporate baffles.

***Slurry Control.*** The most prevalent controls are the measurement of the initial ingredients and the viscosity of the slurry. These can be supported by tests to determine: pH value, temperature, specific gravity, and percentage $SiO_2$ [8].

## Ceramic Shell Production

***Shell Build-Up.*** The pattern assembly is dipped into the primary slurry and rotated to ensure complete contact between the patterns and the slurry. The assembly is then withdrawn and the excess slurry is allowed to drain back into the tank. To ensure the development of an even coating the assembly is normally rotated, usually in more than one orientation, during the draining stage. Immediately after draining the assembly is stuccoed, either in a raining cabinet or fluidised bed. In the former, particles are allowed to fall onto the rotating assembly, whereas in the latter, the rotating assembly is dipped into a tank containing particles which behave like a fluid under the action of pressurised air rising from the porous base-plate of the tank.

After stuccoing, the binder is gelled either by chemical means or by drying in air. In the latter case, drying is accomplished by the circulation of air of controlled temperature and humidity. Careful control of these parameters is important if shell cracking, caused by the temperature differential between the wax and the drying air, is to be avoided. Hardening by chemical gelation reduces the time between coats and therefore improves productivity, but shells that are gelled without drying are generally weaker, even when subsequently dried [8]. Figure 7.2 shows the automated production of ceramic shell moulds.

***Dewaxing.*** Once the required shell thickness has been established the wax pattern can be removed. This is a critical stage, as the expansion of the wax during heating can generate stresses which are sufficiently high enough to crack the green shell. To ensure the highest possible green strength, shells are dried for an extended period of time, 16 to 48 hours, before dewaxing [8]. To minimise the problem of wax expansion and shell cracking, the shells are heated from the outside, causing the surface layer of wax to melt before the bulk of the pattern expands under the action of the heat. The molten wax either runs out of the mould or soaks into it, thus providing space for the expansion of the bulk of the pattern. Two methods of dewaxing are practised: flash dewaxing and autoclave dewaxing. In the former method the shell is inserted in a hot furnace, in the inverted position, whereupon the wax melts and drains out, frequently burning in the process. The process is dirty and the high temperature employed can promote shell distortion. The autoclave is the preferred method of dewaxing, although the capital cost of the equipment is high. Superheated steam at a pressure of about 6 bar (86 lbf/in$^2$) is used to provide the necessary rapid heating. The dewaxing cycle is usually completed within 15 minutes and the wax is recoverable for reclamation.

Fig. 7.2 Photograph showing the automated production of ceramic shell moulds

***Shell Firing.*** The dewaxed shells are fired to remove moisture (free and chemically combined), to burn off residual pattern material and any organics used in the shell slurry, to sinter the ceramic, and to preheat the mould to the temperature required for casting [8]. In many cases this is accomplished in a single operation. Alternatively, preheating may be performed as a second firing operation. In between the two operations the moulds can be cooled, inspected and repaired if necessary. Oxidising conditions are recommended to promote the combustion of organic materials and the firing temperature should be within the range 870 to 1095°C (1600 to 2000°F) [8].

## The Properties and Testing of Ceramic Shell Moulds

The strength of a ceramic shell mould is a function of such factors as: mould material, binder, shell build-up procedure, and the firing procedure. Strength is most commonly determined by a flexural strength/modulus of rupture (MOR) test. Values are quoted at ambient temperature and/or a selected elevated temperature, the latter being more indicative of strength under pouring conditions.

The permeability of a ceramic shell mould has an important influence on the mould filling capability of an alloy cast in air. The rapid displacement of air by the incoming metal stream is essential if the metal is to replicate the mould detail. Unfortunately, the use of very fine refractories to promote good surface finish legislates against the attainment of high permeability in the shells. Although permeability characteristics can be altered by modifications to the slurries and stuccoing procedure, the improvement in permeability may be at the expense of shell strength, and resistance to hot mould deformation in particular. Permeability, like strength, may be determined at ambient temperature, but it is the permeability at the elevated temperature associated with pouring which is the significant factor. In an investigation to establish the factors influencing hot permeability, it was found that marked improvements in hot permeability could be obtained by firing shells to a temperature of 1200°C (2200°F). This enhancement in hot permeability was attained without loss of hot shell strength [17].

## Preformed Ceramic Cores

Preformed ceramic cores are usually produced by specialist ceramic manufacturers who have a variety of methods at their disposal [18]. Cores can be produced by pressing from dry powders, or slurry cast with a hardening binder in the slurry formulation. However, the more usual method of forming cores is by an injection process in which the ceramic powders are blended with a thermo-setting or thermo-plastic binder and injected into a steel die. Depending on the binder hardening mechanism, the die may be either hot or cold. After forming the cores are fired to remove temporary binders and provide a sintered bond. A selection of ceramic cores is shown in Fig. 7.3.

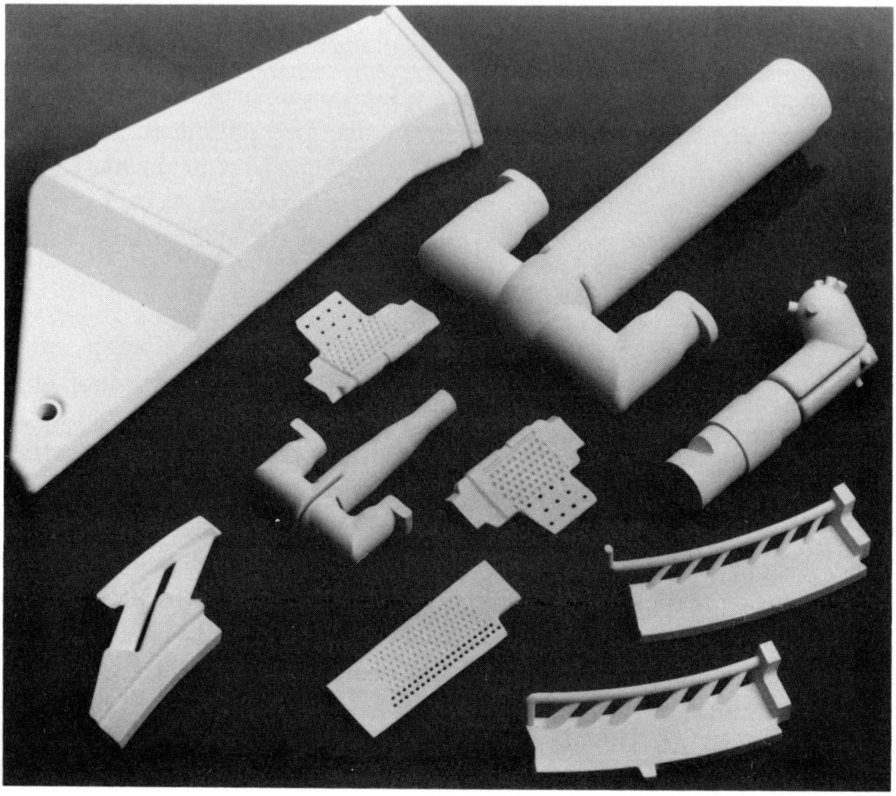

Fig. 7.3 Photograph showing a selection of ceramic cores

Although a considerable number of materials have been proposed for ceramic cores, they are usually made from fused silica, sometimes with additions of zircon. Fused silica cores provide satisfactory refractoriness and are readily leached from the castings in molten caustic baths, aqueous solutions of caustic in open pots or autoclaves, and hydrofluoric acid solutions.

## The Properties and Testing of Ceramic Cores

The desirable properties of a preformed core include [19]:

***Controlled Strength.*** The core should have sufficient strength to avoid breakage during the mould manufacturing and casting stages. However, it should not be so strong that it creates contraction stresses in the solidifying metal which lead to hot tearing or cracking of the casting.

***Refractoriness.*** As the core can reach temperatures above 1000°C (1832°F) in the casting process, poor refractoriness would lead to core softening, resulting in unacceptable dimensional variation in casting wall sections.

***Inertness.*** The core must be stable in air or vacuum throughout its temperature cycle. It must not be affected by, or react with, any of the other process materials. This applies not only to the molten alloy, but also to the mould firing atmosphere, steam in the dewaxing cycle, and the mould materials that hold the core within the mould cavity.

***Ease of Removal.*** With simple core geometries, mechanical removal of the core from the casting may be possible. However, with complex shapes, the core has to be removed by a chemical leaching agent that dissolves the core but does not attack the casting.

***Consistency.*** To ensure consistent quality in casting production, there must be consistency in the cores used. The dimensions, composition and properties of the cores must be reproducible.

## Gating and Feeding

The general principles of good gating and feeding practice should be applied to castings produced by the process. Alloys which are melted and cast under vacuum will not be prone to gas entrapment or oxidation, so that latitude is available in gating system design. However, care must still be exercised when using multiple gating system design for thin section components, to ensure that balanced mould filling is achieved.

## Casting Finishing

The ceramic shell material may be removed from castings by mechanical means, shot blasting, or high-pressure water blasting. These techniques may be successful in removing core material if access is possible. However, it may be necessary to resort to leaching methods. Otherwise, finishing is by conventional means.

## Reclamation

The admixture of different refractory materials and gradings in shell building means that reclamation is inappropriate. Shell building is therefore conducted using materials obtained directly from refractory suppliers.

## Health and Safety Aspects

The comments here are essentially similar to those made in respect of the ceramic moulding processes.

## Process Capabilities

***Alloy Range.*** The process is suitable for a wide range of ferrous, non-ferrous and special alloys.

***Casting Size and Weight Range.*** Whilst the majority of castings produced would be described as small, the size and weight of castings produced by the process has been increasing. Ferrous castings of up to 250 kg (550 lb) finished weight are produced.

***Production Volumes and Lead Times.*** The process requires precision tooling for the production of the wax patterns and the process of producing castings is quite time consuming. These aspects combine to extend the lead time. Expensive tooling usually requires reasonable volumes to recover tooling costs. However, the process is expensive and might, therefore, not be considered for mass production. In practice, the benefits of the process and the absence of an alternative combine in the process being specified for a range of order quantities.

***Thin Section Capability.*** Sections of 1.5 mm (0.060 in) are routinely possible. Thinner sections are possible but depend upon the alloy type and method of processing.

***Detail Replication.*** This is a function of the detail attainable in the pattern and its replication by the mould. The practice of preheating moulds and casting under vacuum or vacuum assistance improves castability and replication of detail. In general, therefore, the process offers excellent detail replication.

***Surface Finish.*** A surface roughness of 3.2 μm (125 μin) is readily attainable. Better values can be obtained but depend upon alloy and processing conditions.

***Cast Holes.*** Holes as small as ø 1 mm (0.040 in) can be cast, providing the length is relatively short.

***Dimensional Accuracy.*** As a general guide, the tolerances for linear dimensions are ± 0.13 mm/25 mm (± 0.005 in/in).

***Typical Castings.*** Investment castings are used in a very wide range of applications which include: aerospace, automotive, computer and electronics applications; valves and fittings; armaments; prosthetic appliances; hand tools; and leisure equipment. Photographs showing a selection of ferrous and non-ferrous castings produced by the process are shown in Figs. 7.4 and 7.5 respectively.

Fig. 7.4 Photograph showing a selection of ferrous castings produced by the investment casting process

Fig. 7.5 Photograph showing a selection of non-ferrous castings produced by the investment casting process

## COUNTER GRAVITY INVESTMENT CASTING PROCESSES

The counter gravity processes, as applied to investment casting, were developed by Chandley and Lamb at the Hitchiner Manufacturing Company during the early 1970's and patented in 1975 [20]. The acronyms CLA and CLV, originally derived from the names of the inventors, are widely used to define the two principal process variants [21]:

CLA counter-gravity low-pressure casting of air-melted alloys
CLV counter-gravity low-pressure casting of vacuum-melted alloys

### Process Descriptions

***CLA Process.*** In this process the hot ceramic shell mould is secured in a specially constructed chamber with its extended sprue protruding downwards and through the base of the chamber, see Fig. 7.6. The chamber, which is sealed from the atmosphere, is lowered towards the furnace crucible so that the extended sprue enters the metal. Vacuum suction is then applied to the chamber, causing the metal to be drawn into the mould.

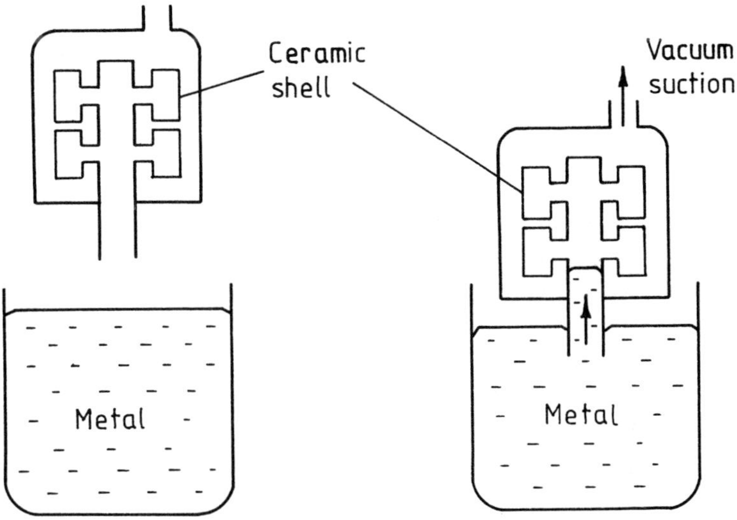


Fig. 7.6 Schematic representation of the CLA process

***CLV Process.*** In this variant the chamber containing the mould is positioned above the chamber of a vacuum melting furnace. A valve and sealing arrangement connects the two chambers, see Fig. 7.7. The metal is melted under vacuum and when it is ready for casting the chamber containing the mould is evacuated. Both chambers are then back-filled with argon gas. At this stage the valve between the chambers is opened and the crucible is elevated until the sprue penetrates below the metal surface. Vacuum suction is then applied to the chamber containing the mould, causing the metal to be drawn into the mould.

Conventional investment casting shell building methods are used and there is nothing exceptional about the requirements for shell strength and permeability [22]. However, a different approach to gate and sprue design is required. The gate size is determined by the need to ensure that it solidifies as quickly as possible after the casting has solidified. This enables the vacuum to be released and any liquid remaining in the sprue to fall back into the crucible. In this respect the process is most suited to the production of thin section castings which will solidify quickly. This minimises the cycle time. If thick section castings were processed in this way solidification time would be delayed and, consequently, productivity reduced. However, a method has been devised to overcome this problem and is termed check valve casting, designated CV. This adaptation requires a flexible fill pipe which can be crimped shut after the mould has been filled [21]. The mould can be extracted from the furnace and transferred to a holding area for solidification to be completed.

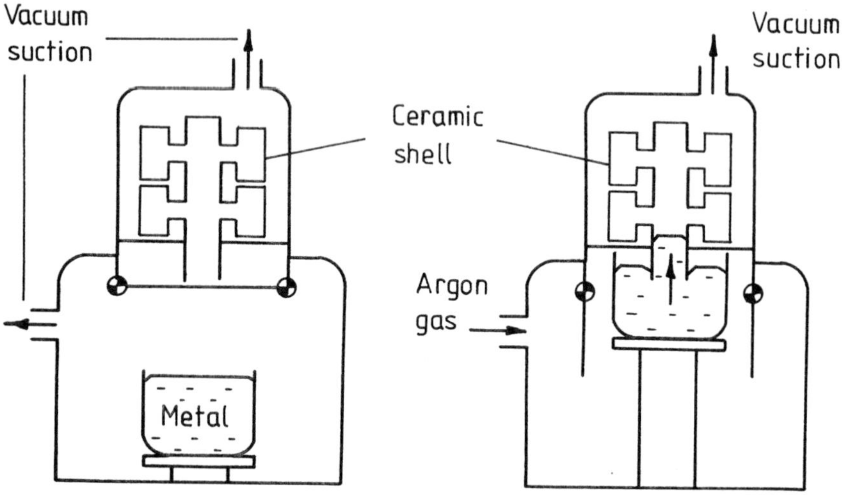


Fig. 7.7 Schematic representation of the CLV process

It is claimed that these vacuum-assisted counter-gravity processes provide improvements in metallurgical quality beyond those attainable in pressure-assisted counter-gravity processes [23]. Both types of process significantly improve metallurgical integrity when compared with conventional gravity pouring methods. Top poured metal is highly susceptible to turbulence, oxidation and slag entrapment. In consequence, the castings contain inclusions. The CLA and CLV processes overcome these problems and therefore reduce the presence of inclusions. Pouring temperature can be reduced without impairing metal fluidity, with reductions in the range of 75 to 150°C (170 to 300°F) claimed [23, 24]. Moulds are placed under less strain, which means that distortion is less evident and flatness is maintained [24].

The advantages of the process are summarised below [22, 24, 25]:

- high yield (up to 90%).
- improved productivity.
- consistent mould filling.
- inclusion free castings.
- lower pouring temperature.

- section thickness of 0.75 mm (0.03 in) routinely possible.
- finer grain size.
- reduced casting distortion.
- cleaner castings.
- reduced finishing costs.

It is not claimed that the processes are suitable for all close-tolerance components, but they are especially suited to the production of thin section castings and those required to be net or near net shape [24].

**Process Capabilities**

***Alloy Range.*** The use of an inert, ceramic shell mould means that the process is suitable for a wide range of alloys. However, it has been applied primarily to ferrous alloys and superalloys.

***Casting Size and Weight Range.*** These are similar to those for the investment casting process. However, there is a size or volume constraint imposed by the size of the mould containers.

***Production Volumes and Lead Times.*** These are similar to those for the investment casting process.

***Thin Section Capability.*** Whilst it is claimed that sections of 0.75 mm (0.03 in) are routinely possible, the generally attainable minimum thickness will be dependent upon the metal being cast.

***Cast Holes.*** No data available.

***Surface Finish.*** Depending upon the alloy being cast, values of surface roughness between 1.6 and 3.2 μm (63 and 125 μin) are readily attained by the processes.

***Dimensional Accuracy.*** Attainable accuracy, in relation to size of dimension, is shown in Table 7.3.

Table 7.3 *Attainable Dimensional Accuracy in Castings Produced by the CLA/CLV Processes*

| Dimension Size | | Tolerance | |
|---|---|---|---|
| mm | in | mm | in |
| up to 6 | up to 0.25 | ±0.08 | ±0.003 |
| 6 to 12 | 0.25 to 0.50 | ±0.10 | ±0.004 |
| 12 to 25 | 0.50 to 1.00 | ±0.125 | ±0.005 |
| over 25 | over 1.00 | ±0.125 | ±0.005 |

***Typical Castings.*** Castings produced by the CLA process for the automotive industry include: steering system components, transmission parts, diesel precombustion chambers, and rocker arms; for the aircraft/aerospace industries: temperature probes, fuel pump impellers, missile wings, brake parts, pump housings, and structural parts. Other applications include: valves and fittings, machine parts, tools, and gun parts. The CLV process is used for processing superalloy parts such as: burner cans, cast airfoils, turbine seals, conduits, clamps, and turbocharger wheels [21].

## THE REPLICAST CERAMIC SHELL (CS) PROCESS

The evaporative pattern casting (EPC) process and the investment casting process share the common feature of the absence of a mould joint line, which reduces the finishing requirement on the castings produced. The Replicast CS process represents the successful development by SCRATA of an alternative process which confers the advantages of both of these processes in the production of precision castings [26]. The major differences between the conventional EPC process and the Replicast CS process are that the expendable foam pattern is coated with ceramic, using the technique of successive dipping and stuccoing employed in investment casting, and is removed before the metal is poured.

Whilst the investment casting process is acknowledged to be pre-eminent in the production of precision castings, it has its limitations [27]. In general, the process is limited to the production of relatively small castings, although larger castings with thin sections can be produced. These limitations arise because of problems associated with shrinkage, distortion and fracture of the wax patterns during manufacture and handling, problems which increase with size and/or section thickness. Furthermore, to avoid cracking during dewaxing the ceramic shell must be relatively thick and, consequently, is expensive.

The Replicast CS process overcomes these problems through the use of a cellular foam, such as expanded polystyrene (EPS), for the pattern. This enables large, lightweight patterns of any realistic section thickness to be produced at reasonable cost. As this foam pattern collapses under the action of heat the ceramic shell strength requirement is reduced and, consequently, its thickness and cost can be reduced.

It is the elimination of the pattern before casting which differentiates the Replicast CS process from the conventional EPC process. Whereas the production of a pattern of minimum density and therefore minimum gas evolution is a principal objective in the conventional EPC process, this is not an objective when the pattern is to be removed before casting. Consequently, the pattern density is determined solely by the requirement to produce a pattern of the highest surface quality, which is achieved at a high pattern density [27]. Patterns produced in high density EPS are structurally stronger, enabling larger patterns to be produced.

The shell build-up procedure is similar to that employed in investment casting. In the original development a slurry based on fine molochite and an ethyl silicate binder were used for the dip coat, although flow coating was used to overcome the problem of pattern buoyancy [26]. The coated pattern was then stuccoed with coarser molochite and each layer was gelled in an ammoniating cabinet. This approach can be varied and one North American foundry dips the EPS patterns in a slurry of ethyl silicate and a blend of fused silica and zircon flour [28]. A primary stucco of -100 mesh (AFS) fused silica is applied by raining and the assembly air-dried for one hour prior to ammoniating. Subsequent stuccoing is performed using a relatively coarse alumino-silicate refractory, the grade of which depends upon the casting size and complexity.

Typically, between three and five coats are applied to the pattern to produce a shell thickness of between 3 and 5 mm (0.125 and 0.188 in). This shell thickness is quite acceptable for castings in the weight range of 50 to 100 kg (110 to 220 lb) [28-30]. The mould assembly is fired to around 1000°C (1832°F), which eliminates the pattern and produces an inert shell. The shell is allowed to cool before casting as it is claimed that the thin shell reduces metal chilling tendency and the likelihood of mould fracture, associated with the high temperature gradient in thick shells, is reduced [30]. The thin shells must be supported during pouring and this is achieved by bedding them in dry, unbonded sand. This sand must be vibrated, using advised procedures, to ensure adequate mould support and the avoidance of mould failure [27]. The support can be enhanced by the use of vacuum assistance.

It is not claimed that the process will provide all the benefits of the investment casting process, especially with regard to surface finish and dimensional accuracy [30]. However, the process may be of interest to investment casters seeking to extend the weight/size range of their products. It certainly provides conventional steelfounders with a process for producing castings of improved quality with reduced finishing costs. An example of a ceramic shell mould produced by the process is shown in Fig. 7.8.

Fig. 7.8 Photograph showing a lightweight ceramic shell produced by the Replicast CS process. In the background a polystyrene pattern is receiving its primary ceramic coating.

## Process Capabilities

***Alloy Range.*** As the mould is inert, the process is applicable to a wide range of alloys. However, it is used predominantly in the production of steel castings and, unlike the conventional EPC process, is suitable for low-alloy and stainless steel casting production because there is no danger of carbon pick-up.

***Casting Size and Weight Range.*** The process is suitable for steel castings in the weight range up to 100 kg (220 lb), which encompasses the majority of steel castings production.

***Production Volumes and Lead Times.*** The process requires the manufacture of precision tools for the manufacture of foam patterns. Reasonable production runs will be required to cover the tooling cost. Whilst the production of the mould can be accomplished more quickly than normal for investment casting, the time taken to produce the pattern will be longer.

***Thin Section Capability.*** Information on this aspect is not available. No claims are made for the process in this respect. Whilst the use of thin shells and vacuum assistance will clearly improve castability, this may be offset by the absence of mould preheat.

***Cast Holes.*** Through holes can be cast providing that the diameter exceeds 9.5 mm (0.375 in).

***Surface Finish.*** Values of surface roughness between 1.6 and 3.2 μm (63 and 125 μin) are readily attainable for steel castings produced by the process.

***Dimensional Accuracy.*** Attainable accuracy, in relation to size of dimension, is shown in Table 7.4 [28].

Table 7.4 *Attainable Dimensional Accuracy in Castings Produced by the Replicast CS Process*

| Dimension | | Tolerance | |
|---|---|---|---|
| mm | in | mm | in |
| 1.5 to 100 | 0.1 to 4.0 | ± 0.25 | ± 0.010 |
| 101 to 305 | 4.1 to 12.0 | ± 0.76 | ± 0.030 |
| 306 to 610 | 12.1 to 24.0 | ± 1.52 | ± 0.060 |

***Typical Castings.*** Valve castings, including those for the petrochemical industry, drive sprockets, truck suspension supports, castings for the food processing industry, and castings for military applications are produced by the process. Figure 7.9 shows steel castings produced by the process.

Fig. 7.9 Photograph showing steel castings produced by the Replicast CS process

## PLASTER MOULD INVESTMENT CASTING

In this process conventional wax patterns are invested with plaster of Paris to produce a plaster block mould. Careful control is exercised over the selection and blending of fillers and additives to the plaster to ensure consistency in the mould material. Production of the block moulds may be conducted under vacuum conditions to eliminate air and ensure that the plaster replicates the pattern surface without blemishes such as small air bubbles. The moulds must be dewaxed and then stored to ensure that all free and combined moisture is removed prior to casting. Although some permeability can be introduced to the block mould by the incorporation of additives which burn out during stoving, the permeability is still relatively low. To overcome this problem, vacuum assistance may be used to aid filling of the mould at the casting stage.

## References

1. Taylor, P. R. (1986). Lost wax casting - a short illustrated review. *Metals and Materials,* **2**, 11, 705-710.

2. Krohn, B. R. (1984). The art and science of investment casting. *Modern Casting,* **74**, 12, 22-26.

3. Smart, R. F. (1989). High strength and integrity investment castings. *Materials and Design*, **10**, 6, 301-305.

4. Smart, R. F. (1989). The investment casting market. *Foundry Trade Journal*, **163**, 3397, 678, 680, 682.

5. Russell Stratton, R. (1988). The 1988 investment casting industry: today and beyond. *Proceedings of the Conference on High Integrity Castings.* Published by ASM International, 35-43.

6. Williams, R. G. and Smart, R. F. (1990). Market trends in the investment casting industry. *Paper presented to the Annual Conference of the Institute of British Foundryman*, 12 pp.

7. Anon. (1990). The investment casting market. *Foundry Trade Journal,* **164**, 3413, 325-327.

8. Horton, R. A. (1988). Investment casting. In: *Metals Handbook, Vol. 15 Casting*, 253-269. Published by ASM.

9. Mitchell, S. R. (1988). Wax - an historical pattern for tomorrow's process. *Foundry Trade Journal,* **162**, 3379, 875-876.

10. Williams, R. B. (1988). Update on investment casting waxes. *Proceedings of the 7th World Conference on Investment Casting*, Paper No. 2, 8 pp.

11. Chambers, C. A. (1985). Understanding waxes. *Proceedings of a one-day Seminar on Wax Pattern Making.* BICTA, 1.1-1.5.

12. Carpenter, L. (1988). The use of speciality waxes in the investment casting process. *Proceedings of a Symposium on Investment Casting Waxes.* AFS/CMI, 91-98.

13. The BICTA Patternmaking Committee (1985). *The Manufacture and Use of Soluble Wax Cores.* Published by BICTA, 9 pp plus appendices.

14. Argueso, L. (1986). History of pattern waxes composition, characteristics and handling. *Steelfounder's Res. J.*, **15**, 15-18.

15. Phipps, B. (1988). Wax injection equipment. *Proceedings of a Symposium on Investment Castings Waxes.* AFS/CMI, 28-35.

16. *Wax, Binder and Slurry Test Handbook.* Published by the American Foundrymen's Society, Des Plaines, Illinois, USA, 1988.

17. Lang, W., Learmouth, D. and McCallum, R. (1988). Hot permeability of shell moulds: its measurement and place in process technology. *Proceedings of the 7th World Investment Casting Conference.* Paper No. 3, 25 pp.

18. The BICTA Mould Materials Committee (1972). *Preformed Ceramic Cores: Available Manufacturing Methods.* Published by BICTA, 6 pp.

19. Mills, D. (1979). Properties of ceramic cores. *Proceedings of the 15th Annual BICTA Conference*, Paper No. 7, 14 pp.

20. Chandley, G. D. (1976). Making castings without ladles or sprues - the CLA process. *Trans. AFS,* **84**, 37-42.

21. Chandley, G. D. (1988). Counter gravity low-pressure casting. In: *Metals Handbook, Vol. 15 Casting*, 317-319. Published by ASM.

22. Anon. (1975). Hitchiner's new investment casting process. *Foundry,* **103**, 10, 26-28.

23. Chandley, G. D. (1989). Quality enhancement by fluid flow control in counter gravity casting processes. *Cast Metals,* **2**, 1, 2-10.

24. Taylor, I. C. (1988). CLA process for thin-wall castings. *Castings Buyer,* **1**, 3, 12-13.

25. Recaldin, A. J. (1985). Cost effective investment casting of thin sections - the CLA process. *Design Engineering*, **10**, 149, 153, 157.

26. Bish, D. A. and Hubbard, P. A. (1982). Application of ceramic coatings to expendable patterns. *Proceedings of the SCRATA Conference,* Paper No. 17, 6 pp.

27. Ashton, M. C., Sharman, S. G. and Brookes, A. J. (1984). The Replicast CS (ceramic shell) process. *SCRATA Journal of Research*, **64**, 1-11.

28. Grote, R. E. (1986). Replicast ceramic shell molding: the process and its capabilities. *Trans. AFS*, **94**, 181-186.

29. Brown, J. R. (1984). Applications of the REPLICAST full-mould and ceramic shell process. *Proceedings of the BCIRA Conference*, Paper No. 27, 14 pp.

30. Hubbard, P., Brookes, A. J., Sharman, S. G. and Ashton, M. C. (1983). The Replicast CS (ceramic shell) process. *Proceedings of the SCRATA Conference*, Paper No. 3, 8 pp.

CHAPTER 8

# The Die Casting Processes

## INTRODUCTION

The die casting processes employ a permanent, reusable mould or die, generally made of metal, and this feature distinguishes them from the refractory aggregate moulding processes considered so far. The die casting processes are precision casting processes and so must be included in a treatise on the subject. However, as this text was written to complement titles on pressure die casting in the publisher's list [1, 2], it is intended only to outline the processes, their capabilities and recent developments. The four processes considered are:

Gravity Die Casting (permanent moulding)
Pressure Die Casting (die casting)
Low Pressure Die Casting
Squeeze Casting

Through their use of permanent metal dies these processes are characterised by relatively high tooling costs combined with low production costs. The processes are suited to the mass production of castings, predominantly in the non-ferrous alloys. This restriction is imposed by the reduction in die life which results when high temperature alloys are processed in metal dies. In terms of volume of output, aluminium alloys are the most important and are suited to each of the four processes considered. Zinc alloys are produced by pressure die casting and the process is also used in the production of magnesium- and copper-base alloys. Ferrous materials can be die cast, grey cast iron is processed successfully by gravity die casting, and steel components have been produced by pressure die casting and squeeze casting. Melting temperature and die life considerations apart, not all alloys are suitable for die casting. To be suitable for die casting an alloy must not be susceptible to cracking due to the constraint imposed by the die during solidification and contraction of the casting.

Capital costs, in terms of equipment and dies, can be very high for these processes and this is the principal reason for the requirement for large order quantities. These costs are generally at their highest for pressure die casting and squeeze casting and lowest for gravity die casting.

## GRAVITY DIE CASTING

The gravity die casting or permanent mould casting process is quite simple in concept. A permanent, usually metal, mould replaces the expendable, refractory aggregate mould of sand and similar processes. Molten metal is poured into the permanent mould or die where it remains until it has solidified. The mould or die, which is usually made in two halves, is then opened and the casting is removed. The process is particularly suitable for the medium to high volume production of castings (predominantly non-ferrous) with fairly uniform wall thickness and simple shape. Whilst there are similarities between the sand casting and gravity die casting processes, the latter permits the production of more uniform castings with closer dimensional tolerances, superior surface finish and improved mechanical properties [3]. Simple cores, made of metal, or complex cores, made of sand or plaster, can be used with the process to improve the degree of casting complexity attainable.

The advantages of the process may be summarised as follows:

- castings can be produced to a higher level of consistency than can sand castings.
- gravity die castings generally have a better metallurgical structure and higher mechanical properties than sand castings.
- the surface finish of gravity die castings is generally better than that for sand castings.
- castings produced by the process may require less machining than sand castings.
- the porosity level of gravity die castings is generally lower than that of sand castings.

The disadvantages of the process include:

- higher tooling costs requiring larger order quantities.
- not all alloys are suitable for casting by the process.
- the process is not as flexible as sand casting in terms of design complexity, size and weight range.

## Tooling Requirements

The dies can be of relatively simple construction, produced in two halves and jointed vertically, see Fig. 8.1.

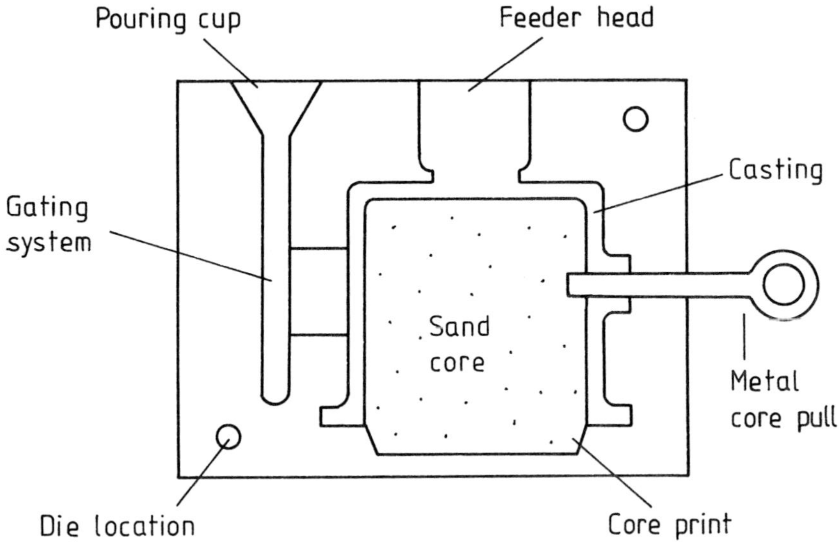


Fig. 8.1 Schematic diagram of one half of a gravity die casting die

The dies must contain provision for accurate location, clamping, gating, feeding, venting and casting ejection. The majority of dies are manufactured from cast iron, which means that tooling costs and lead times are significantly lower than those for pressure die casting. Low carbon steels and die steels provide improved thermal shock resistance, but not necessarily improved erosion resistance, and may be preferred for higher melting point casting alloys. However, cast iron castings are produced from grey iron dies. Graphite moulds may be used for small production runs of most metals and have been used successfully for medium volume zinc-base alloy castings production [4].

Although expensive materials, beryllium-copper alloys provide good die life when used for copper-base alloys [5].

Die life varies with casting design, die material and the casting alloy, but could exceed 250,000 pours for components such as petrol or diesel engine pistons [3]. Die design is important to die life and a uniform wall thickness, typically 30 to 45 mm (1.2 to 1.8 in) [6], is recommended to minimise differential expansion/ contraction, which can lead to warping and eventually thermal fatigue failure. Dies incorporate an allowance for contraction and generous taper is advocated to minimise cracking or sticking of the casting during contraction. Magnesium-base alloys are especially prone to cracking and a taper of between 5 and 10° is advised.

The dies are coated with a refractory layer which serves several purposes. Firstly, it insulates the die and reduces premature chilling of the metal, and secondly, it provides a barrier to direct contact between the die surface and the casting alloy, thus reducing the tendency to erosion. Lubricant coatings may be sprayed over insulating coatings. They usually contain graphite and they are applied to the die to facilitate casting removal.

## Casting Production

In order to ensure that castings of consistent quality are produced by the process, attention must be paid to the following factors: melt quality, temperature control, gating, and feeding.

***Melt Quality.*** To ensure optimum mechanical properties melts should be adequately degassed, inclusion-free and subjected to close temperature control. In these respects the requirements are no different from those for other precision casting processes.

***Temperature Control.*** The temperature of the molten metal and that of the die have a direct effect on the quality of the casting, productivity and die life. Close control over both aspects is therefore extremely important.

***Gating.*** Good gating design should minimise turbulence, which will cause air entrapment or oxidation of the metal, and also the velocity of the metal stream at the gates, to minimise erosion. Tapered sprues which are kept full during pouring reduce the problems of turbulence.

***Feeding.*** Metal dies promote progressive solidification, so that castings in long freezing range alloys will be sounder when produced in gravity dies rather than sand moulds. However, the provision of feed metal will be necessary and, as in sand casting, feeder heads must be provided according to the general rules of feeding.

**Process Developments**

Developments in the gravity die casting process have been directed toward the mechanisation of the process to minimise operator involvement and simultaneously improve die life, product consistency, quality, and productivity. An example of a simple mechanised gravity die casting machine is shown in Fig. 8.2.

Fig. 8.2 Photograph showing a simple mechanised gravity die casting operation

Where high productivity is required, multi-station carousels may be employed with as many as twenty die sets/carousel. Hydraulic actuation permits such actions as die opening/closing, die clamping and casting ejection to be performed automatically. Improved product quality can be achieved by the controlled, automatic dispensing of metal and by mould rotation to facilitate die filling. Thermal control of the dies using water channels or sequential water spray and air cool reduces cycle time. However, this must be practised with care, otherwise accelerated die failure by thermal fatigue may be a consequence. Die temperature-cycle time controllers are used for sequencing die opening and closing to optimise both rate of production and casting quality [7].

Counter-gravity casting, using pressure to displace metal from a holding furnace upwards into a die cavity, is the principle adopted in the low pressure die casting process discussed later in this chapter. However, the technique of vacuum suction to draw the metal upwards from a holding furnace into the die cavity is applied to gravity die casting [8]. It is claimed that the use of vacuum improves the quality and mechanical properties of gravity die castings.

**Process Capabilities**

***Alloy Range.*** Although more castings are produced in aluminium-base alloys, there is significant production of copper-base alloys (including brass, gunmetal, aluminium and manganese bronzes), cast iron, magnesium-base alloys, and zinc-base alloys.

***Casting Size and Weight Range.*** For general production, casting weights would not exceed:

70 kg (155 lb) for aluminium-base
25 kg (55 lb) for magnesium-base
13.6 kg (30 lb) for cast iron
9 kg (20 lb) for copper-base alloys [3]

***Production Volumes and Lead Times.*** Cast to shape tooling produced in grey cast iron reduces the time and cost of tooling when compared with pressure die casting, but lead time and cost will be higher than for sand casting. Reasonable order quantities are required to justify the cost of tooling.

***Thin Section Capability.*** The process is not noted for its thin section capability and 5 mm (0.20 in) would be a sensible minimum.

***Detail Replication.*** The process is not noted for detail replication and is not as good as sand casting in this respect.

***Surface Finish.*** The surface finish is in part dependent upon the metal being cast, but typically will be in the range of 3.2 to 6.3 μm (125 to 250 μin).

***Cast Holes.*** As a general guide the minimum diameter for cored holes is 6.5 mm (0.25 in).

***Dimensional Accuracy.*** Dimensions may be held to within ± 0.25 mm (± 0.01 in) in a single die half, but critical dimensions will require machining.

***Typical Castings.*** Castings in aluminium-base alloys include inlet manifolds, cylinder heads, brake master cylinders, and pistons for automotive applications. Castings in copper-base alloys include gear selector forks in aluminium bronze.

## PRESSURE DIE CASTING

The pressure die casting process has its origins in typecasting machines, which had reached a high level of automation and mechanical efficiency by the mid-1800's [9]. By the end of the 19th century variants of the typecasting machines were being used to produce components for cash registers and gramophones, and by the beginning of the current century die cast bearings were being produced for automotive applications [1]. Although the technology of the process continued to evolve during the early part of the 20th century, it was not until the 1920's that the forerunners to the modern hot chamber and cold chamber machines were developed. These processes have been refined to the extent that tiny zinc alloy castings can be produced in a one second cycle on fully automated hot chamber machines. The cold chamber process is used predominantly for the production of aluminium alloy components, which include complex castings weighing in the region of 15 kg (30 lb), for the automotive industry. For such components the cycle time would be about two minutes.

### Process Outline

***Hot Chamber Process.*** A schematic diagram of a hot chamber pressure die casting machine is shown in Fig. 8.3. The metal for casting is maintained at an appropriate temperature in a holding furnace adjacent to, if not part of, the machine. The injection mechanism is located within the holding furnace and a substantial part of it is therefore in constant contact with the molten metal. Pressure is transmitted to the metal by the injection piston, which forces it through the gooseneck and into the die. On the return stroke metal is drawn into the gooseneck for the next shot. In this process there is minimum contact between air and the metal to be injected, thus minimising the tendency for turbulent entrainment of air in the metal during injection. However, there is prolonged contact between the metal and parts of the injection system, which effectively restricts this process to zinc-base alloys.

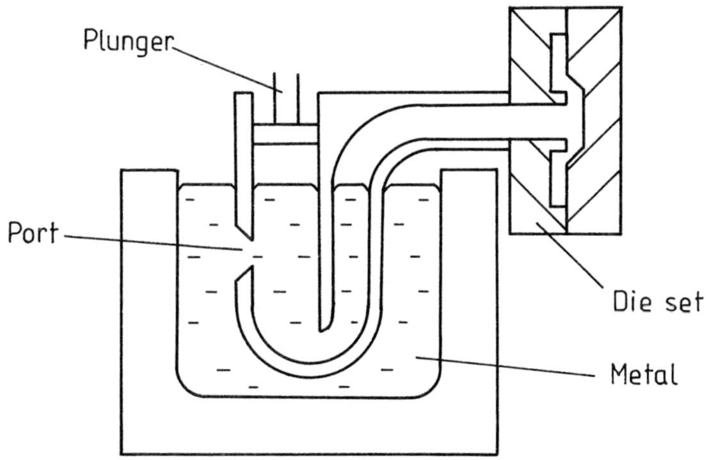


Fig. 8.3 Schematic diagram of a hot chamber pressure die casting machine

***Cold Chamber Process.*** A schematic diagram of a cold chamber pressure die casting machine is shown in Fig. 8.4.

The essential feature of this process is the independent holding and injection units. In the cold chamber process metal is transferred by ladle, manually or automatically, to the shot sleeve. Actuation of the injection piston forces the metal into the die. This is a single-shot operation. This procedure minimises the contact time between the hot metal and the injector components, thus extending their operating life. However, the turbulence associated with high speed injection is likely to entrain air in the metal, which can cause gas porosity in the castings. The cold chamber process is used for the production of aluminium- and copper-base alloys and has been extended to the production of steel castings. Figure 8.5 shows a die for the production of castings by the cold chamber process.

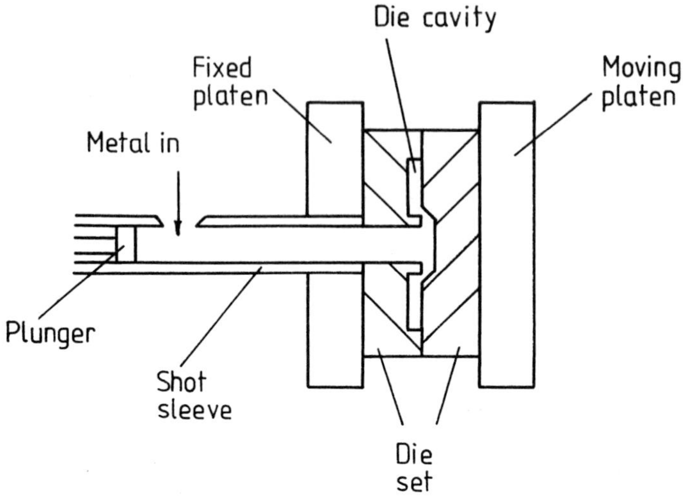


Fig. 8.4 Schematic diagram of a cold chamber pressure die casting machine

The advantages of the pressure die casting processes may be summarised as follows:

- the ability to produce castings with close dimensional control.
- the ability to produce castings with a good surface finish.
- the ability to produce castings with thin walls, and therefore of reduced weight.
- the ability to produce castings at a high rate of production.

Against these advantages the following disadvantages should be weighed:

- high capital plant costs.
- high tooling costs.
- restrictions on the range of alloys which can be cast.
- restrictions on the maximum size of casting that can be cast.

Fig. 8.5 Photograph showing castings extracted from a cold chamber pressure die casting machine

**Tooling Requirements**

The dies for pressure die casting are frequently complex assemblies which are precision machined in heat resistant steels. They are expensive pieces of equipment. The dies themselves are often mounted in bolsters which locate directly onto the platens of the die casting machine. The bolsters may be produced in cheaper materials because they are not directly exposed to the molten metal. Die design must incorporate provisions for: gating, location of die halves, core actuation, casting ejection, venting, and die cooling. The assembly of a die from several components increases its complexity and cost but provides greater flexibility in the casting design. Multi-part dies also provide the opportunity to replace parts which are subject to excessive wear without the cost associated with replacement of the whole die. To maximise production from a given die multiple impressions are used.

The factor with the greatest influence on die life is the metal pouring temperature, as shown in Table 8.1. Dies fail primarily as a result of heat checking, which is a form of thermal fatigue resulting from the rapid changes in temperature during the casting and cooling cycles.

Table 8.1 *Typical Die Life when Casting Various Alloys [10]*

| Alloy Type | Die Life (number of shots) | Pouring Temperatures (°C) | (°F) |
|---|---|---|---|
| Brass | 20,000 | 950 | (1740) |
| Aluminium-Base | 100,000 | 700 | (1290) |
| Magnesium-Base | 300,000 | 670 | (1210) |
| Zinc-Base | 1,000,000 | 420 | (790) |

**Casting Production**

The principal variables to be controlled during the production of a casting by the pressure die casting process are: die temperature, metal pouring temperature, injection pressure, and injection speed.

***Die Temperature.*** Close control over the die temperature is very important in the maintenance of product quality. Die temperature has a direct influence on such factors as surface finish, dimensional accuracy and internal soundness. Under normal operating conditions an equilibrium value is reached and maintained during continuous operation. However, stoppages for any reason will lead to fluctuations in die temperature and, in consequence, casting quality. A die temperature control facility is therefore an essential feature of die casting equipment.

***Pouring Temperature.*** Variations in pouring temperature will directly affect the cycle time and adversely affect the attainment of equilibrium conditions. In addition, too high a pouring temperature will increase the amount of shrinkage present in the casting and reduce die life, too low a temperature will increase the likelihood of misrun castings. Adequate arrangements must therefore be made for control of the melt temperature, together with its cleanliness and gas content.

***Injection Pressure.*** The most complex injection systems are found on cold chamber machines. The filling of dies can be considered to occur in three stages. During each stage the injection piston will move at a different speed. In the first stage, or slow approach stage, the piston moves forward to displace the metal in the shot sleeve in such a way that air is expelled through the die venting system. The removal of the air in the shot sleeve is essential if the casting is to be produced with the minimum of air entrainment. The second stage is a fast stage during which metal is injected into the die cavity as quickly as possible to ensure complete die filling. Once the die is full the pressure builds up to its highest level. Pressure intensification may be employed at this stage to minimise the presence of shrinkage or gas porosity in the resultant casting. A final pressure under these circumstances would generally be between 50 and 70 MPa (3 and 4.5 tonf/in$^2$).

***Injection Speed.*** Die filling times in the order of 0.05 to 0.15 seconds are common. It is not possible to recommend the ideal fill time, which is a compromise between a fast fill time, required to optimise surface finish, and a slow fill time, required to minimise air entrapment. Whilst there is clearly a relationship between fill time and metal velocity, the two are not synonymous. The latter is a function of gate area and has, at high values, a direct influence on die erosion. As a guide, gate velocities should not exceed: 40 m/s (130 ft/s).

The high injection pressures used in the pressure die casting process require correspondingly high locking forces to keep the dies closed. The locking force required is a function of the injection pressure multiplied by the projected area of the casting in the plane normal to the die closing direction. Machines are usually described in terms of their locking force in tonnes (tons) and several machine manufacturers produce machines rated at 3500 tonnes (3445 tons) and above.

The advent of the microprocessor and low cost computing has enabled manufacturers to instrument their equipment so that close control can be exercised over the process variables. This enables casting producers to improve their control over casting quality and provides customers with a very high level of consistency in the castings they purchase.

## Process Capabilities

***Alloy Range.*** Zinc-base alloys produced by the hot chamber process, and aluminium-base alloys produced by the cold chamber process, are the alloys most suited to the pressure die casting process. However, some copper-base alloys and an increasing quantity of magnesium-base alloys are also produced by the proccss.

***Casting Size and Weight Range.*** Most production is in the size and weight range which would be described as small. However, the size and weight range for aluminium alloy castings continues to increase to satisfy the demands of the automotive industry.

***Production Volumes and Lead Times.*** Complex precision tooling requires a relatively long lead time for its production and its cost determines that significant production volumes are required. However, production runs as low as 5000 may be justifiable. The rate of output ranges from 1 shot/second for tiny zinc alloy castings to 15 shots/hour for large aluminium alloy automotive castings.

***Thin Section Capability.*** Sections as thin as 0.5 mm (0.02 in) in zinc-base alloys, 0.8 mm (0.03 in) in aluminium-base alloys and 1.5 mm (0.06 in) in copper-base alloys can be produced by the process.

***Detail Replication.*** The use of pressure ensures that the metal replicates the die detail with a high degree of fidelity.

***Cast Holes.*** Holes can be cast to a very high standard of accuracy: ± 0.015 mm (± 0.0006 in).

***Dimensional Accuracy.*** Critical dimensions below 25 mm (1 in) in size can be held to ± 0.08 mm (± 0.003 in) for zinc-base alloys, ± 0.10 mm (± 0.004 in) for aluminium-base alloys and ± 0.18 mm (± 0.007 in) for copper-base alloys. For dimensions between 25 and 300 mm (1 and 12 in) an extra allowance for each 25 mm (1 in) increase of ± 0.025 mm (± 0.001 in), ± 0.038 mm (± 0.0015 in) and ± 0.051 mm (± 0.002 in) should be allowed respectively for the three alloy groups. An additional tolerance will be necessary for dimensions formed across the die joint line or produced by moving die parts.

***Surface Finish.*** A good standard of surface finish is attainable on castings produced by the processes, better for the alloys poured at lower temperatures, and can be expected to be in the range 0.4 to 3.2 μm (16 to 125 μin).

***Typical Castings.*** A very wide range of castings is produced by pressure die casting and includes, for automotive applications: carburettor bodies, distributor bodies, housings for clutches and gearboxes; for electrical applications: motor frames and cases, switchgear housings and covers; for general engineering applications: pulleys and rotating parts, portable tool covers, record player parts, etc.

**Process Developments**

Pressure die castings are noted for their precision in terms of dimensional accuracy, surface finish, detail replication, and thin sections. However, these external qualities were not generally matched by metallurgical integrity. As a consequence of turbulent filling, gases were entrapped in the metal. Apart from their direct effect on the reduction in mechanical properties, these gases precluded post-casting processes which involved the application of heat because the heat expanded the gases, causing blistering on the surface of the casting. Furthermore, porosity arising from shrinkage is difficult to overcome because of the complexity of the castings and the constraint imposed by solidifying metal, which prevents the injection pressure from acting to promote the flow of still liquid metal for feeding purposes.

Although the problem of turbulence has been largely overcome by improvements in gating system design and injection control, there still remains the need to ensure soundness in those castings required for critical applications. This is especially the case when heat treatment is necessary to develop the optimum properties in the alloy. Developments which have the objective of improving the metallurgical integrity of die castings are considered in this section.

***The Acurad Process.*** To improve the soundness of hypereutectic aluminium-silicon alloy cylinder block castings for the Vega engine, General Motors developed the Acurad process. The name Acurad deriving from the objectives of the programme, which were to produce castings accurately, rapidly and to an improved standard of density [11]. The main features of the process are:

- controlled cooling of the die
- large ingate
- lower metal injection velocity or rate
- use of a secondary plunger

Die filling is achieved at a comparatively slow rate and low injection pressure through a comparatively large gate, and these factors combine to minimise turbulence and the associated gas entrapment. The unique feature of the process is employed once the cavity has been filled. The process employs two concentric plungers which advance together during the die filling stage. As the casting begins to solidify it does so primarily at the interface with the die. Once this casting skin has formed it restricts the movement of the external plunger. However, the internal or secondary plunger remains free to exert pressure on the remaining liquid, thus providing a degree of pressure assisted feeding. The larger gate employed ensures that the gate does not freeze before the secondary pressure is applied. The principle of the process is shown in Fig. 8.6.

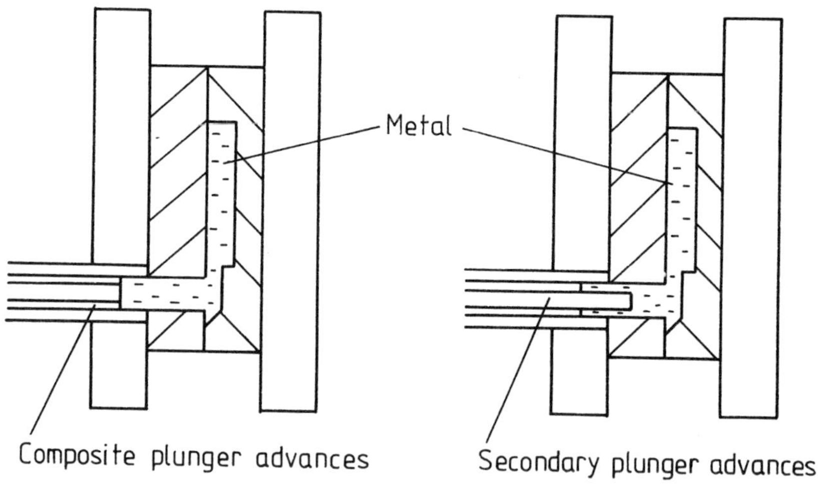


Fig. 8.6 Schematic diagram showing the principle of the Acurad process

It is claimed that the process produces castings with a density some 3 to 5% higher than conventional die castings [12]. The advantages resulting from the use of the Acurad process include castings with improved pressure tight integrity that are amenable to welding, heat treatment and coating without blistering [13].

***Pore-Free Die Casting.*** If it is assumed that there are no gases dissolved in the melt, then the gases entrapped within the metal at the die filling stage must arise from the air in the cavity and gases which are the product of die coating/ lubricant breakdown. Scientists at ILZRO analysed the gas entrapped in zinc die castings and found that the main element present was nitrogen [14]. The conspicuous absence of oxygen suggested that it was consumed during the casting process, reacting to form zinc oxide in zinc die castings or aluminium oxide in aluminium die castings. On this basis, if the die cavity were to be flushed with oxygen prior to the injection of metal, then the nitrogen, as the major source of gas porosity, would be displaced from the die through the venting system and a casting free from gas porosity should result. The oxygen-purge is the principle upon which the pore-free die casting process was developed.

In the process oxygen is introduced through the shot sleeve with the die halves slightly apart, to provide venting of the cavity air, and the plunger advanced to cover the pouring hole. After the purging cycle the die is closed and casting proceeds in the conventional manner [15]. The process is slightly slower than conventional pressure die casting, due to the purging cycle. Special die lubricants are required, water-based with colloidal graphite, and this reduces the life of the plunger and shot sleeve. As the oxygen is consumed within an enclosed space there is a pressure drop within the cavity which enhances the die filling capability [14]. The oxide produced is very fine, typically 1 μm (40 μin) in size and in total the oxide accounts for less than 0.2% of the casting weight [16]. To maximise the benefits of the process the die must be designed to minimise shrinkage and, in the case of aluminium alloys, the melt must be properly degassed to minimise the presence of hydrogen porosity.

The advantages claimed for the process over conventional pressure die casting include [16]:

- gas porosity is virtually absent.
- castings can be heat treated.
- castings can be used at elevated temperature.
- castings can be used in welded assemblies.
- thinner and lighter castings can be produced, since there is no gas to displace from the die.
- mechanical properties such as tensile strength and ductility are improved.
- improvements in fatigue properties are obtained.

Castings produced by the process in aluminium alloys include: car wheels, motor cycle wheels, brake master cylinders, suspension arms, air-conditioner compressor components, hydraulic power steering cases, bicycle hubs and crank cases, heat exchangers, and agricultural equipment [16].

***Vacuum Die Casting.*** As its name implies, the vacuum die casting process attempts to minimise gas porosity by reducing the amount of air in the die cavity before metal injection commences. No special seals, gaskets or heads are required and die design, construction and operation is in most respects unchanged [17]. However, precautions must be taken to ensure that the molten metal is not drawn into the vacuum system. One method relies on a chill-vent which is a water-cooled, corrugated, thin sectioned vent that quickly freezes any metal that enters it, thus preventing it from reaching the vacuum pump [17]. Other methods, such as the Hodler Optivac system [18] and the Ube Gas Free system [19], use cut-off valves activated by the metal pressure to prevent the metal entering the vacuum pump.

***The Vacural Process.*** In this process vacuum suction is used to draw the metal from the holding furnace into the shot sleeve [20]. It is, therefore, quite different in its concept to the vacuum die casting process described in the preceding section. The method of metal transfer is claimed to reduce the presence of inclusions in the castings. The application of a vacuum during casting enables lower injection pressures and therefore smaller machines to be used [18].

***Secondary Pressurisation.*** The problem of shrinkage, especially that in remote sections such as bosses, may be overcome by secondary pressurisation, which has a similar objective but different method of application to that employed in the Acurad process. In secondary pressurisation, also known as the pressure pin or additional pressure exertion process, pins - not unlike ejector pins - are actuated so that they penetrate the casting surface. By exerting local pressure they provide pressure assisted feeding and an improvement in soundness. These pins are usually driven into areas on the casting which are drilled out afterwards [15].

***The Vertical Pressure Die Casting Process.*** The concept of vertical die casting is not new, as such a machine was developed during the 1920's [21]. However, the machine developed by Toyota and Ube, to complement their squeeze casting machine development, has features designed to overcome the problems associated with conventional pressure die casting [22]. Its objectives are to prevent the inclusion of aluminium oxide in the metal, to avoid turbulent occlusion of air and to provide pressure assisted feeding. The principle of the machine is shown schematically in Fig. 8.7.

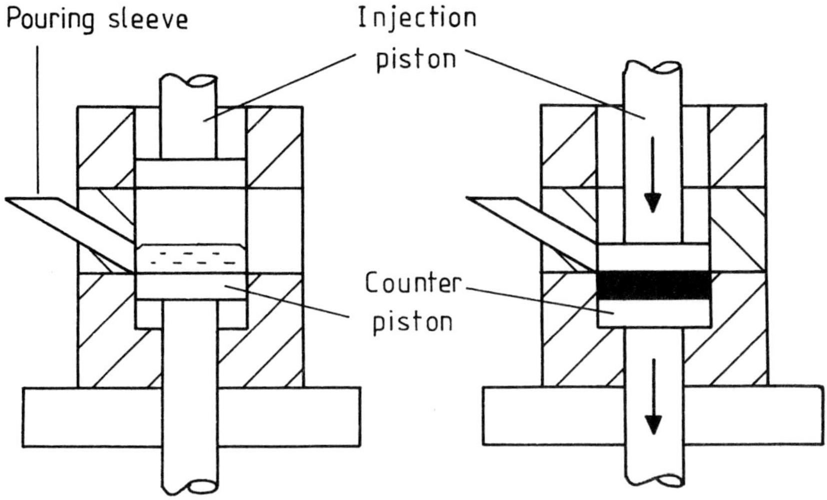


Fig. 8.7 Schematic diagram of a vertical pressure die casting machine

The molten metal is introduced to the shot sleeve through a pouring sleeve. The shot sleeve is sealed at its base by the counter piston. The injection piston is actuated in the vertical downward direction and, concurrently, the counter piston moves in the same direction. Metal can now enter the gates to the radially distributed cavities, which it does in a quiescent manner. The pressure is maintained at a level above 30 MPa (2 tonf/in$^2$) to ensure pressure assisted feeding of the castings.

It is claimed that the method of cavity fill does not enclose gas in the metal, that any aluminium oxide remains on the surface of the metal adjacent to the injection piston and that solidification shrinkage is eliminated. The castings can be heat treated and typical castings produced by the process include a power steering component, a valve rocker arm and pistons incorporating a fibre preform located at the wear ring groove.

***The Counter-Pressure Casting Process.*** Although this process has features which are similar to low pressure die casting and the counter gravity casting processes, it has one feature which is unique, and that is that at the beginning of the operating cycle the whole system is pressurised [23]. The die is contained within an upper pressure chamber which sits upon a lower pressure chamber that contains the molten metal. The two are connected by a fill tube which penetrates below the surface of the metal, see Fig. 8.8.

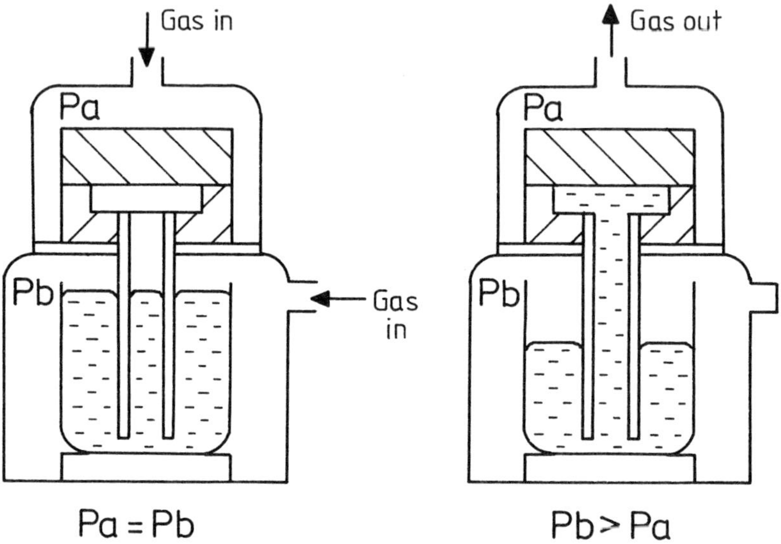


Fig. 8.8 Schematic diagram of the Counter-Pressure casting process

To effect die filling a pressure difference is established between the two chambers so that metal rises into the die at a controllable and adjustable speed. Turbulence is eliminated and the pressure ensures good replication of the die shape by the casting. The maintenance of pressure above the melt prevents the evolution of dissolved gas, and shrinkage porosity is eliminated through the provision of risers subjected to continuous pressure.

***Rheocasting.*** The Fulmer Research Institute claims that not only can air entrapment in pressure die castings be reduced but shrinkage porosity may be reduced, yield and die life increased, and die design simplified if the rheocasting concept is adopted [24]. By vigorously agitating an alloy as it cools through the region between its liquidus and solidus, a slurry is produced in which the dendrites have been broken down to form globular solid particles. The semi-solid alloy may be processed by pressure die casting, or a related process, to produce a shaped casting or, alternatively, processed through a continuous casting unit which is the Fulmer approach.

Once agitation ceases, the solid particles in the slurry begin to agglomerate and the viscosity of the alloy increases. However, shearing will break the bonds between the agglomerates and the apparent solid will flow like a liquid. If the alloy is allowed to solidify without shearing after agitation ceases, the material retains its unique structure so that, if it is re-heated to between the solidus and liquidus, shearing will once more enable the alloy to flow.

Material produced by the Fulmer route can therefore be used as bar feedstock, cut to a predetermined weight and re-heated, to provide an individual charge for the pressure die casting operation. Injection of the apparently solid charge through the ingate provides the shearing action, filling is achieved without the turbulence associated with the injection of superheated liquid, and the associated lower temperature improves die life. As the injected alloy is partially solid there is reduced solidification shrinkage to accommodate and less latent heat of fusion to dissipate. Whilst it is not claimed that shrinkage porosity is entirely eliminated, the absence of entrained air permits heat treatment of the castings to realise high mechanical properties. Fulmer claim that components produced by this route have performed satisfactorily in domestic equipment, cars and other vehicles [25].

***Ferrous Die Casting.*** Considerable effort and resources have been expended to develop a ferrous die casting process. The GKN Ferrodi process [26] is a cold chamber pressure die casting process based on a conventional machine adapted for ferrous production to incorporate special features in the metal contact areas such as the dies, injection system and metal melting.

Considerable technical problems had to be overcome, particularly those associated with the need to construct economical dies to withstand extremely severe thermal shock, thermal fatigue and erosion by liquid alloys at temperatures up to 1600°C (2910°F). A molybdenum-based alloy known as TZM exhibits high resistance to thermal fatigue or heat checking. This is largely attributed to its high thermal conductivity and thermal diffusivity which, combined with a low coefficient of thermal expansion, minimise thermal gradients and hence thermal stresses. It is only inserted in those regions of the die contacted by hot metal because of its high cost. A die life of 10,000 shots, which is about three times the economic minimum die life, is attainable.

Castings in the weight range 0.6 to 3 kg (1.3 to 6.6 lb) and in various materials, including stainless steel, have been produced. Good dimensional accuracy is attainable and although castings may contain some dispersed micro-shrinkage, the improved microstructure results in mechanical properties at least equivalent to conventional casting.

## LOW PRESSURE DIE CASTING

The low pressure die casting process represents an attempt to obtain the benefits of die casting without some of the disadvantages. Although the gravity die casting process employs relatively low-cost tooling and can be accomplished without major capital plant expenditure, it cannot be claimed that it is a method for producing castings requiring minimum machining, with minimum section and optimum metallurgical integrity at high yield.

The pressure die casting process is undeniably without competition where the production of large numbers of thin section castings having good surface detail and finish are required, and has the advantage that machining may be eliminated or at least reduced. However, metallurgical integrity is not a feature of pressure die castings and, melt quality apart, they may exhibit shrinkage porosity and gas porosity, the latter as a result of cavity air and/or die coating degradation products entrained in the injected metal stream. The presence of gas porosity not only reduces mechanical properties directly but also prevents the full properties of heat treatable alloy from being realised, because the gas expands during heating and causes blistering of the casting surface.

A feature of the process is the low pressure displacement of liquid alloy through a refractory stalk connecting the holding furnace unit to the base of the closed die, see Fig. 8.9. Quiescent filling, under controlled pressure from the bottom of the die cavity, eliminates entrained air and the sustained pressure during solidification improves soundness. Castings may be heat treated and the integrity attained is demonstrated by the use of the process for automobile wheels and cylinder head castings.

The advantages claimed for the process include the following [24-26]:

- lower capital cost (compared with pressure die casting).
- improved process control (compared with gravity die casting).

- improved metallurgical integrity through reduced shrinkage porosity and non-metallic inclusions.
- good surface definition in castings.
- consistent dimensional accuracy.
- disposable cores may be used with the process.
- high yields attainable.
- suitable for the production of thin-wall castings.

Against these advantages there are a number of disadvantages, relative to the pressure die casting process:

- production rate is slower.
- surface finish is not as good.
- thin section capability is not as good.

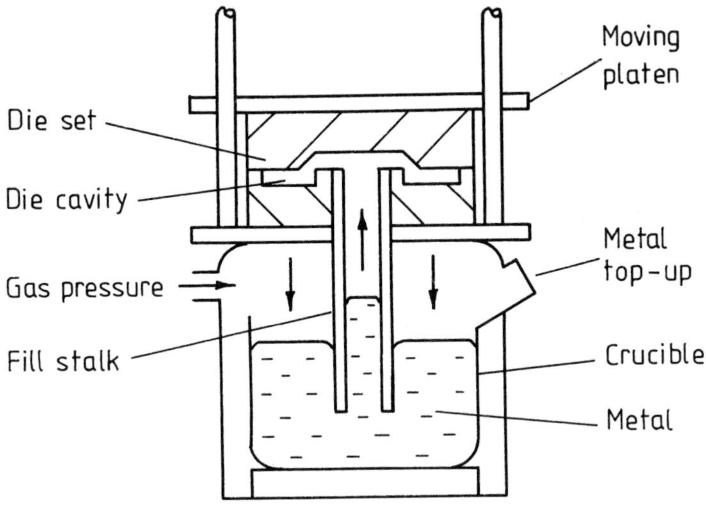


Fig. 8.9 Schematic diagram of the low pressure die casting process

Although the concept of low pressure die casting was developed in the early part of the century, it was not developed to any extent for the production of aluminium alloy castings until World War II. It was not until the mid-1950's that the process was adopted for the production of automotive castings.

## Tooling Requirements

The dies are normally manufactured from cast iron as the pressures involved are very low. The material is comparatively cheap and easy to machine. For high production runs of complex castings, tool steel inserts may be used in the dies. It would also be normal practice to use tool steel for those parts of the die subject to wear. Die life figures of 50,000 shots are typical for the process [27]. Careful attention must be given to the incorporation of vents within the die design and to the provision of casting ejection facilities. A general taper of 2 to 3° is acceptable. As a general guide, tooling costs will be between a third and a half of those for a comparable pressure die casting die [28].

## Casting Production

The low pressure die casting machine can be considered to comprise a metal holding unit, a die set, a die actuation system, and a metal displacement system. Metal for casting is melted in a separate furnace and transferred to the machine's holding unit on demand. The holding unit has its own heating facility to ensure that the metal is kept at a temperature suitable for casting. A fill-tube passes through the pressure cover and lower platen of the machine and extends well down into the metal in the holding unit. As the tube is in constant contact with molten metal at its upper end, it must be heat resistant, resistant to attack by the molten metal, have high thermal shock resistance, and be pressure tight [29]. Cast iron, properly coated with refractory at regular intervals to avoid dissolution of the iron, is generally used. However, refractories such as silicon nitride meet the requirements, although they may be more expensive [25]. The metal is displaced up the tube by introducing air, or an inert gas, into the holding unit, which must therefore be hermetically sealed, at the pressure of 0.25 to 1 bar (3.5 to 15 lbf/in$^2$). The metal enters the die cavity through a gate which is typically ø 20 to 30 mm (ø 0.8 to 1.2 in), regardless of the casting size. It is essential that the dimensions of the gate should be sufficiently large to ensure that the gate solidifies after the casting. The metal enters through the base of the die cavity, displacing the air through the vent system. Die coatings based on the use of fillers such as talc or graphite in a sodium silicate carrier are used to protect the die surface. On completion of die filling, positive pressure is maintained on the metal until the casting has solidified. This requirement has a major influence on the cycle time. Once solidification is complete the pressure is reduced, whereupon the metal in the fill tube returns to the holding unit and the die can be opened for removal of the casting.

The in-line design shown in Fig. 8.10 is the most widely used.

Fig. 8.10 Photograph showing a low pressure die casting operation

However, some installations feature laterally displaced holding units and electromagnetic pumps have been proposed for the dispensation of the metal [30]. Although the process is used predominantly for the casting of aluminium-base alloys, other alloys could be processed by this method. With appropriate process technology, the benefits of low pressure die casting have been demonstrated for magnesium-base alloys [31].

**Process Capabilities**

***Alloy Range.*** The process is used predominantly for the production of aluminium-base alloy castings.

***Casting Size and Weight Range.*** The process is suitable for castings in the weight range of 5 to 25 kg (11 to 55 lb), although castings outside this range are processed economically by the process.

***Production Volumes and Lead Times.*** Tooling production times are likely to approach those for pressure die casting dies. In general, a minimum order requirement of 5000 castings will be necessary to justify the tooling costs. Production output is quite low, for an automotive wheel 15 castings/hour/machine would be considered a high rate of production.

***Thin Section Capability.*** For general production the minimum section recommendation is between 3.5 and 4.5 mm (0.14 to 0.18 in), although thinner sections can be cast.

***Cast Holes.*** A minimum diameter of 6 mm (0.25 in) is specified for holes produced by retractable core pins.

***Detail Replication.*** Detail replication is good, but not quite as good as pressure die casting.

***Surface Finish.*** Values intermediate between those of gravity and pressure die casting can be attained. As a general guide, these will be in the range 1.6 to 6.3 μm (63 to 250 μin).

***Dimensional Accuracy.*** As a general guide, a normal commercial tolerance would be 0.35 mm (0.014 in) on dimensions up to 25 mm (1 in), with a further tolerance of 0.08 mm (0.003 in) for each additional 25 mm (1 in). Dimensions formed across a joint line would require a larger allowance.

***Typical Castings.*** These include: automotive wheels and cylinder heads, gearbox and clutch covers, transmission and differential housings, cylinder heads and blocks for all types of small engines, electric motor stators, transformer covers, and heat sinks.

## SQUEEZE CASTING

Squeeze casting is a single step process for converting liquid metal into a fully shaped component of precise dimensions, excellent surface finish and with metallurgical integrity equivalent to that obtained in forgings. By combining precise metal metering, quiescent die filling and high applied pressure during solidification, air entrapment, oxidation, hydrogen gas nucleation and shrinkage are eliminated. The combination of liquid metal and a vertical movement press suggests a hybrid combination of casting and forging, which has led to the terms liquid metal forging and squeeze forming being used. Other names by which the process is called include: extrusion casting, liquid pressing and pressure crystallisation. In the simplest form of operation, liquid metal is metered into the bottom half die mounted on the lower platen of the press, see Fig. 8.11 a) and b), and the upper, or moving, platen holding the upper die half, or punch, is actuated.

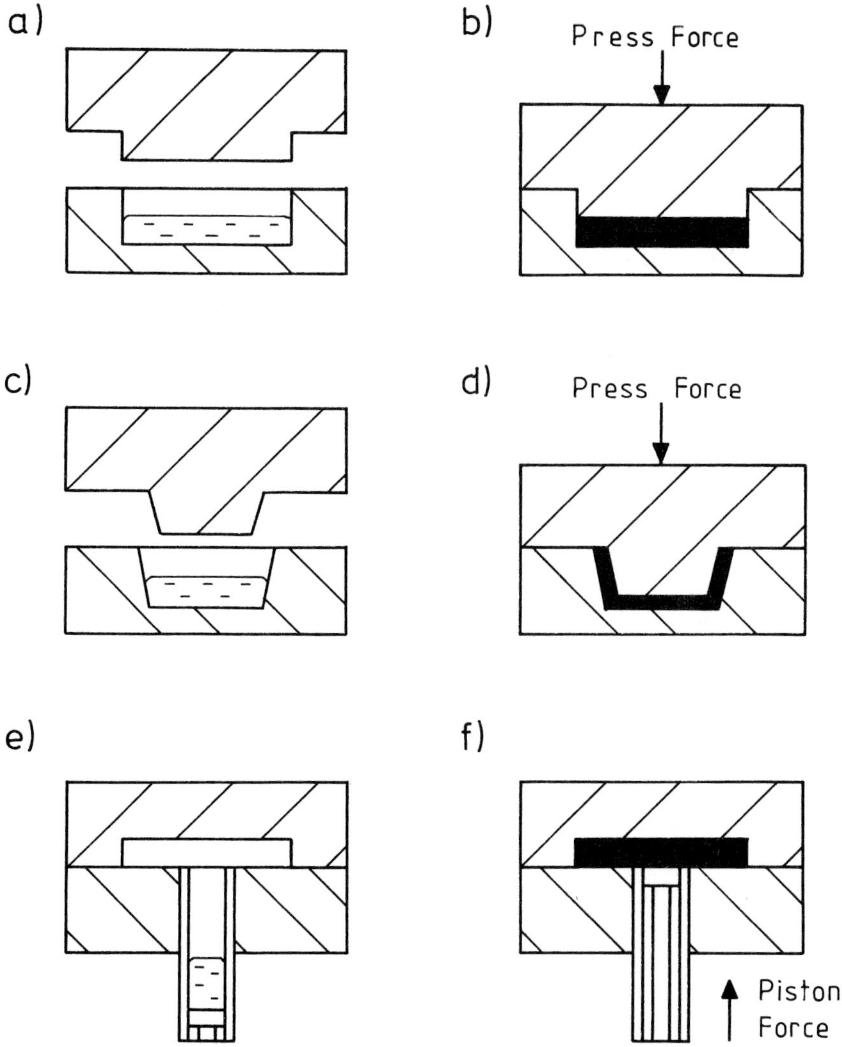


Fig. 8.11 Schematic diagram of the squeeze casting process variations: a) and b) direct pressure application; c) and d) extrusion variant; e) and f) indirect pressure application

Closing of the die halves is accompanied by transmission of the press force through the punch on to the liquid metal. Not only does the sustained pressure during solidification eliminate shrinkage, it also ensures intimate contact between the metal and die, which accelerates solidification, improves the metallurgical structure and promotes precision. Close control of such variables as: metal temperature, die temperature, pressure, pressure delay time, and pressure application time is essential for consistent operation.

Two variations of the squeeze casting theme are shown schematically in Figs. 8.11 c) and d) and 8.11 e) and f). In the extrusion variant, the liquid metal is extruded backwards around the contoured punch, enabling a more complicated casting shape to be obtained. The indirect pressure variant is essentially a vertical die casting arrangement which features a controlled slow fill followed by high applied pressure. For further details of the process, reference may be made to reviews by Rajagopal [32] and Franklin and Das [33]. Advantages claimed for the process include the following [32, 33]:

- ability to produce parts with complex profile and thin sections beyond the capability of conventional casting and forging techniques.
- substantial improvement in material yield because of the elimination of gating and feeding systems.
- significant reduction in pressure requirements, in comparison with conventional forging, while at the same time increasing the degree of complexity that can be obtained in the parts.
- ability to use both cast and wrought compositions.
- improvements in product quality in regard to surface finish, dimensional accuracy and mechanical properties.
- castings may be heat treated.
- complete elimination of shrinkage and/or gas porosity.
- potential for using cheaper, recycled material without the loss of properties that would occur with other processes.
- the potential for increased productivity in comparison with gravity die casting or low pressure die casting.
- suitable for the production of cast composite materials.

Against this must be weighed the disadvantages, which include:

- high capital costs, comparable to pressure die casting but lower than forging.
- relatively low die life, in comparison with the true die casting processes.

- shape complexity limitation, in comparison with the true die casting processes.
- An accurate metal metering system is required.

**Tooling Requirements**

Dies for squeeze casting must be of a sufficiently rigid construction to withstand the pressures applied. They are usually constructed from high quality die steels such as H13, which is generally specified for aluminium alloy casting production. Extended die life for ferrous or copper-base alloys may be obtained by using refractory metal (molybdenum or tungsten) alloys. Minimum corner radii of 2 mm (0.08 in) for non-ferrous alloys and 3 mm (0.12 in) for ferrous alloys are recommended. A taper of 0.5° on the die and between 1 and 2° on the punch is recommended to facilitate casting extraction. Adequate venting of the dies is important and can be achieved in part at the joint plane, through joints, or by the incorporation of machined vents. Vents should be less than 0.2 mm (0.008 in) to avoid melt extrusion and excessive flash production. A die coating is generally employed to protect and lubricate the die surfaces in contact with liquid metal. Coatings based on graphite are recommended for non-ferrous alloys and ceramic coatings for ferrous alloys. Multiple cavity dies can be used providing that the press has sufficient power and multi-part dies may be employed in order to increase shape complexity.

**Process Parameters**

The metal supplied to the squeeze casting die should be free from oxides and preferably low in gas content. Non-turbulent melting and transfer procedures are desirable and filtration is a viable option. To ensure consistency in operation, and especially dimensional accuracy, metal metering is practised. The principal variables in the process, which require optimisation for each alloy type and casting configuration, are: pouring temperature, die temperature, applied pressure, and pressure application time.

***Pouring Temperature.*** The pouring temperature has a direct influence on both the casting quality and die life. If the pouring temperature is too low then cold laps and premature solidification are likely to occur. If the pouring temperature is too high then extrusion from the die through vents is possible, together with the formation of porosity in thicker sections. Die life will be adversely affected by high pouring temperatures. The pouring temperature will depend upon the alloy, but typically will be between 6 and 55°C (45 and 130°F) above the liquidus [34]. In general, short freezing range alloys require a high pouring temperature to avoid solidification of the metal before the pressure is applied. A low pouring temperature is more effective in yielding good metallurgical quality in long freezing range alloys, providing that the applied pressure is sufficiently high to eliminate porosity [35].

***Die Temperature.*** Temperatures within the range of 200 to 400°C (392 to 752°F) are quoted for the process, with the range of 200 to 300°C (392 to 540°F) being most often specified for aluminium-base alloys. To ensure adequate clearance for venting, the punch temperature should be between 15 and 30°C (27 and 86°F) below that of the die [34].

***Applied Pressure.*** The level of applied pressure depends upon the alloy being cast, the casting geometry and the required level of mechanical properties. For most alloys there is an optimum pressure level beyond which there will be no further improvement in mechanical properties. Pressures in the range 30 to 140 MPa (2 to 9 tonf/in$^2$) are most often quoted [34, 36]. This pressure must be maintained throughout the solidification period for the casting.

***Pressure Application Time.*** As indicated, the pressure must be maintained throughout the solidification period. Times between 30 and 120 seconds are quoted but will be dependent upon the part geometry. Maintaining pressure beyond solidification simply increases the cycle time. The cycle time for a squeeze casting is generally shorter than that for an equivalent casting produced by either gravity die casting or low pressure die casting. This is because the rate of heat transfer is so much higher in the squeeze casting process.

### Metallurgical Characteristics

Squeeze castings are characterised by their metallurgical integrity, achieved through the application of sustained high pressure during solidification. The absence of voids in the casting ensures that the mechanical properties will be enhanced. The microstructure of squeeze cast material will be affected by the rapid rate of solidification arising from the contact maintained between the metal and the die as a result of the application of pressure during solidification. Refinement of the microstructure has been identified as contributing to the improvement in mechanical properties of several alloy systems processed by the squeeze casting route. Improvements in proof stress, ultimate tensile strength and ductility, in both the as-cast and heat-treated conditions, have been reported. Wrought alloy compositions may also be processed successfully by the squeeze casting route as the requirement for 'castability' is of less importance in the process.

Casting defects can arise in squeeze castings, the most common ones being: oxide inclusions, porosity, extrusion segregation, centreline segregation, blistering, underfill, cold laps, hot tearing, sticking, case debonding, and extrusion. The cause of these defects and the methods for their control are described by Rajagopal and Altergott [37].

## Cast Metal Matrix Composites

The squeeze casting process has featured prominently in the development of cast metal matrix composite, i.e. metal matrix composite (MMC), material. Its principal advantage in this application is the production of a defect-free matrix, which is critical to the attainment of improved mechanical properties. In composite production by squeeze casting two approaches may be used. In the first, the composite alloy is produced by stirring the reinforcement into the liquid alloy. The resultant material can then be shaped by the pressure applied by squeeze casting. In this approach the reinforcement is distributed throughout the casting. If a more discriminating incorporation of reinforcement is required, for example to produce the required properties in a specific area of the casting, then a preform can be inserted in the die, with application of pressure during casting ensuring infiltration of the preform by the metal. MMC materials provide improved properties which include: improved strength, better retention of strength at elevated temperatures, improved modulus (stiffness), potential for component weight reduction through improved strength to weight ratio, improved wear resistance, and reduced coefficient of thermal expansion.

## Indirect Squeeze Casting

The original Ube vertical squeeze casting machine was developed jointly with the Toyota Motor Corporation and was first used for the production of cast automobile wheels in 1976. The features which distinguish this machine from the direct pressure machine are the patented tilting-docking mechanism of the shot unit and the controlled slow-filling of the die by bottom injection followed by sustained pressure [38]. The non-turbulent transfer of metal and very low gate velocity, compared with pressure die casting, eliminates the problem of trapped air in thicker section castings. To enable thin section castings to be produced a higher gate force is necessary. To overcome the air entrapment problem that would accompany this higher velocity, a vacuum system is employed. Shrinkage porosity is eliminated by the maintenance of pressure between 50 and 90 MPa (3 and 6 tonf/in$^2$) on the casting during solidification. The original vertical shot, vertical clamping design is used for the production of pistons, disc drives, connecting rods, compressor housings, and other parts. The more recently developed horizontally clamping, vertical shot machine is used in the production of thin section components such as master brake cylinders, brake calipers, manifolds, fuel rails, valve bodies, gear lever forks, and other components.

**Process Capabilities**

***Alloy Range.*** Most commercial production is of castings in aluminium-base alloys. However, magnesium-, zinc- and copper-base alloys can be processed and feasibility studies have been conducted for ductile iron, steels and superalloys.

***Casting Size and Weight Range.*** The limiting factor will be the size of the press required. In the UK military vehicle wheels weighing 19 kg (42 lb) are in regular commercial production on a press rated at 1500 tonne (1476 ton).

***Production Volumes and Lead Times.*** Expensive precision tooling is required, with an associated time delay in its manufacture. Large production quantities are required to justify the tooling costs but production rates will be relatively slow, at between 15 and 60 castings/hour/machine.

***Thin Section Capabilities.*** Whilst it is claimed that sections as thin as 0.3 mm (0.012 in) have been produced by squeeze casting, the process is better suited to castings with sections above 6 mm (0.24 in).

***Cast Holes.*** Holes can be incorporated in the casting through the use of appropriate die features.

***Detail Replication.*** The application of pressure during casting ensures a high degree of detail replication from the die to the casting.

***Surface Finish.*** The surface finish of squeeze castings is equivalent to that attained by pressure die casting, 0.4 to 3.2 μm (16 to 125 μin).

***Dimensional Accuracy.*** Aluminium alloy castings can be produced to a tolerance of 0.2 mm/100 mm (0.002 in/in).

***Typical Castings.*** Direct squeeze casting is probably most noted for its production of pistons and military vehicle wheels. Other applications include: domestic cookware, truck hubs, brass and bronze bushes and gears. A selection of castings produced by the process is shown in Fig. 8.12.

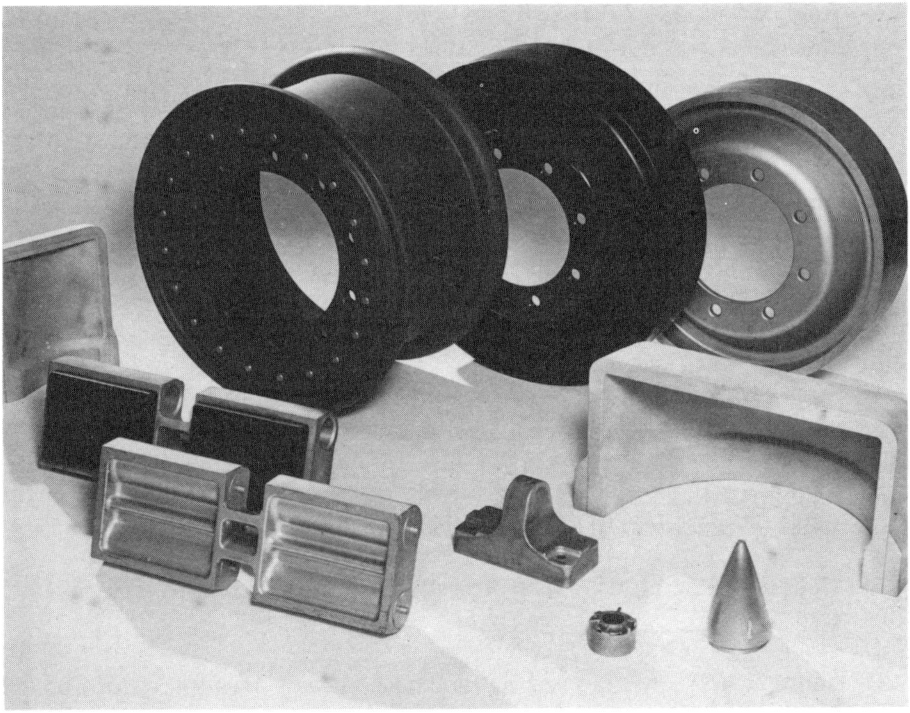

Fig. 8.12 Photograph showing a selection of castings produced by the GKN Squeeze Form process

## References

1. Upton, B. (1982). *Pressure Diecasting Part 1.* Pergamon Press, Oxford.

2. Allsop, D. F. and Kennedy, D. (1983). *Pressure Diecasting Part 2.* Pergamon Press, Oxford.

3. West, C. E. and Gruback, T. E. (1989). Permanent mold casting. In: *Metals Handbook, Vol. 15*, 275-285. Published by ASM.

4. Miahichuk, W. (1986). An introduction to graphite permanent mold casting. *Modern Casting,* **76**, 2, 36-38.

5. Anderson, R. and Michlin, D. (1986). Permanent mold casting of copper-base alloys. *Ibid,* **76**, 11, 40-42.

6. Stahl, G. W. and Whaler, K. R. (1981). Reviewing the permanent mold process, parts 1 and 2. *Ibid,* **71**, 10, 45-29 and 11, 54-58.

7. Kaye, A. (1982). Recent developments in gravity and low-pressure die casting machines. *Foundry Trade Journal,* **152**, 3238, 725-737.

8. Arnold, T. F. (1985). Vacuum improves strength, quality of permanent-mold castings. *Machine Design,* **57**, 13, 83-86.

9. Barton, H. K. (1974). Die casting: past, present and future. *Engineer's Digest,* **35**, 9, 53-59.

10. Birch, J. M. (1988). High pressure die casting of non-ferrous alloys. *Materials Science and Technology*, **4**, 3, 218-221.

11. Bryant, A. L. (1970). Acurad casting of hypereutectic aluminium engine block. *Proceedings of the SDCE Congress,* Paper No. 65, 8 pp.

12. McLaren, J. L. (1966). The Acurad story. *Die Casting Engineer,* **10**, 5, 8-22.

13. Anon. (1969). Acurad casting method updated. *Modern Metals,* **25**, 11, 34.

14. Lapsys, A. L. (1983). Pore free aluminum wheels and other O.E. castings. *SAE Technical Paper 830017*, 7 pp.

15. Booth, S. E., Millgate, R. A., Bonar, J., Harrison, W. G. W. and Clegg, A. J. (1989). Developments in die casting and investment casting in South Korea and Japan. *A report published by the BNF Metals Technology Centre,* 173 pp.

16. Goodwin, F. E. and Hofstatter, A. F. (1983). Pore-free die casting - progress in automotive applications. *Proceedings of the SDCE Congress,* Paper No. 63, 4 pp.

17. Brogden, H. G. (1979). Die design and production criteria when using vacuum in brass die casting. *Ibid*, Paper No. 93, 4 pp.

18. Kaye, A. and Street, A. (1982). *Die Casting Metallurgy.* Published by Butterworth Scientific Ltd, London, 234-235.

19. Sose, M. (1985). Air content and its effect on casting quality. *Proceedings of the SDCE Congress,* Paper No. 34, 5 pp.

20. Zaiss, G. (1985). New steps in die casting technology. *Ibid*, Paper No. 71, 7 pp.

21. Kaye, A. and Street, A. (1982). *Die Casting Metallurgy.* Butterworth Scientific Ltd, London, 5.

22. Tokui, M., Morita, A., Masoaka, T., Ohta, A., Suzuki, I. and Kawano, F. (1983). Development and application of vertical pressure die casting process. *Proceedings of the SDCE Congress,* Paper No. 63, 6 pp.

23. Balevski, A. A. T. (1989). Counter-pressure casting: the processes which occur during casting and crystallisation. *Foundry Trade Journal,* **163,** 3399, 744/6/8-750.

24. Brook, G. B. (1982). Improving the quality of aluminium diecastings by novel techniques. *Materials and Design,* **3**, 5, 558-565.

25. Brook, G. B. (1984). Semisolid metal injection castings. *Proceedings of the Conference on Materials Engineering*, Inst. of Metals, 109-116.

26. Sellers, R. G. R. and Hayes, J. G. (1976). Ferro-Di, GKN's new steel-die casting process, parts I and II. *Foundry Trade Journal,* **141**, Part I 3097, 811-821; Part II 3099, 1024-1030.

27. Seman, N. G. (1979). Low pressure permanent molding: an introduction. *Foundry,* **107**, 11, 46-48/50/52.

28. Plume, A. W. (1983). Low pressure die casting - fact and fiction. *Die Casting Engineer,* **27**, 3, 34-37.

29. Plume, A. W. (1985). Low pressure die casting - the state of the art. *Proceedings of the SDCE Congress,* Paper No. 74, 6 pp.

30. Palmer, J., Howard, P., and Strowbridge, M. (1981). Low pressure die casting. *Engineering, Technical File No. 88,* **221**, i-vii.

31. Strowbridge, M., Williams, P. and Palmer, J. (1979). Low pressure die casting: how, why and when? *Design Engineering,* 10, 121-135.

32. Bryan, P. A. (1972). Low pressure die casting today. *The British Foundryman*, **65**, 11, 401-408.

33. Carbonnel, H. (1976). Low pressure die casting. *Die Casting Engineer,* **20**, 5, 22-24/26.

34. Westengen, H. and Holta, O. (1989). Low pressure permanent mould casting of magnesium - recent developments. *Foundry Trade Journal,* **163**, 3399, 738/40/42.

35. Rajagopal, S. (1981). Squeeze casting: a review and update. *J. Applied Metalworking,* **1**, 4, 3-14.

36. Franklin, J. R. and Das, A. A. (1984). Squeeze casting - a review of the status. *The British Foundryman*, **77**, 3, 150-158.

37. Dorcic, J. L. and Verma, S. K. Squeeze casting. In: *Metals Handbook, Vol. 15. Casting*, 323-327. Published by ASM.

38. Chadwick, G. A. and Yue, T. M. (1989). Principles and applications of squeeze casting. *Metals and Materials,* **5**, 1, 6-12.

39. Barlow, J. (1984). New opportunities via squeeze formed aluminium. *Proceedings of the IAVD Congress on Vehicle Design and Components,* 6 pp.

40. Rajagopal, S. and Altergott, W. H. (1985). Quality control in squeeze casting of aluminum. *Trans. AFS,* **93**, 145-154.

41. Suzuki, S. (1989). Vertical squeeze casting of aluminum components. *Modern Casting*, **78**, 10, 38-40.

CHAPTER 9

# Design for Casting

## INTRODUCTION

The design function is the starting point for all manufacturing activity, including the production of castings. A designer must consider a number of issues at the outset. Firstly, a decision will be taken on what material to specify. This will be determined by the mechanical properties of the material, its density and its suitability for the environmental conditions under which the component must operate. Secondly, the functional shape of the component must be determined in order that it can meet its functional objectives. At this stage the designer will be concerned with such aspects as ease of manufacture, sectional properties and aesthetic appearance.

In choosing the metal casting process as the principal manufacturing route the designer will be seeking to exploit the special versatility that this process provides. To fulfil the designer's requirements the casting must be produced to the required shape, dimensional accuracy and surface finish specified, and be free from surface and internal defects, gross cracks, tears, distortion, and residual stresses.

The designer must be aware of the constraints that may be imposed on a design by the founder's ability to produce thin sections, ensure soundness in remote sections and avoid distortion. A knowledge of metal casting processes, their capabilities and limitations, will therefore be invaluable to the designer, as will the opportunity to liaise with a foundry engineer at the design stage. Liaison at this stage will ensure that the most suitable casting process is selected and that the decision on casting orientation in the mould, which affects such factors as joint plane determination and tooling requirements, ensures cost effective production to the benefit of both the founder and customer alike.

In order to be able to use the metal casting process effectively in component design and production, the designer should be aware of the following considerations [1, 2]:

1. The designer should have a thorough knowledge of the casting processes available, their capabilities and limitations.

2. Casting is particularly suitable for parts which contain internal cavities that are inaccessible, too complex, or too large to be easily produced by machining.

3. Cores should, however, be avoided as far as possible because they reduce the dimensional accuracy of the mould assembly, increase the production cost of the casting and the fettling problems.

4. Complex design details should be modified to simplify production and reduce component costs.

5. The casting process enables the designer to specify walls of varying thickness, bosses, lugs, pads and, strengthening ribs.

6. Thin sections may be desirable, within the limits of strength requirements, for weight saving purposes, but their specification may restrict the choice of casting process.

7. Remote thicker sections and junctions locally increase the casting's mass and present the possibility of shrinkage porosity unless access for feeding is possible.

8. Variations in section thickness result in variations in cooling rate which may, in turn, cause distortion or residual stresses.

9. The specification of close tolerances will increase the cost of the casting and must be weighed carefully against potential savings in the cost of machining.

10. Castings have isotropic properties.

11. Large complicated castings can often be redesigned to be split into several smaller, less complex castings which can be joined subsequently.

## Design for Cost Effective Production

One of the principal reasons behind a designer's choice of the metal casting process is the ability to specify complex shapes. However, complex shapes increase the manufacturing cost and this must be taken into account when designing for cost effective production. The foundry's manufacturing costs will be determined by such factors as casting orientation in the mould, the need for cores and the implication that the casting shape has on fettling requirements. The significance of these factors is most readily demonstrated by examples.

***Casting Orientation in the Mould.*** It is advantageous to orientate a casting so that heavier sections are uppermost, providing good access for feeding. When the replication of fine detail is important, positioning the critical surface low in the mould avoids inclusions and benefits from the increased metallostatic pressure. Orientation directly influences the choice of joint plane and this, in turn, determines the requirement for taper on the pattern, the mould volume, the need for cores, access for gating and feeding, the ease of fettling, and the amount of machining required subsequently. These aspects are demonstrated by the examples shown in Figs. 9.1 to 9.5.

Fig. 9.1a shows a plain cylindrical bush which will be produced as a casting and then machined all over to achieve the final dimensions required by the component. The foundry engineer must decide on the orientation of the casting in the mould before the pattern equipment can be specified. The available options are illustrated in Figs. 9.1b to 9.1e.

If option 9.1b is chosen then the lower mould (drag) half will be quite deep, a large amount of sand will be required to produce the mould and this will increase the production cost. Extensive taper will be required, both internally and externally, to ensure that the pattern will separate readily from the mould. This extensive taper will increase the requirement for machining and consequently the cost to the customer. However, the cost of a core will be saved and the orientation is ideal for directional solidification and top placement of a feeder head. If it is decided that a core should be used to produce the internal diameter then an additional cost arises through the cost of the core, but machining will be reduced as there is no internal pattern taper, see Fig. 9.1c.

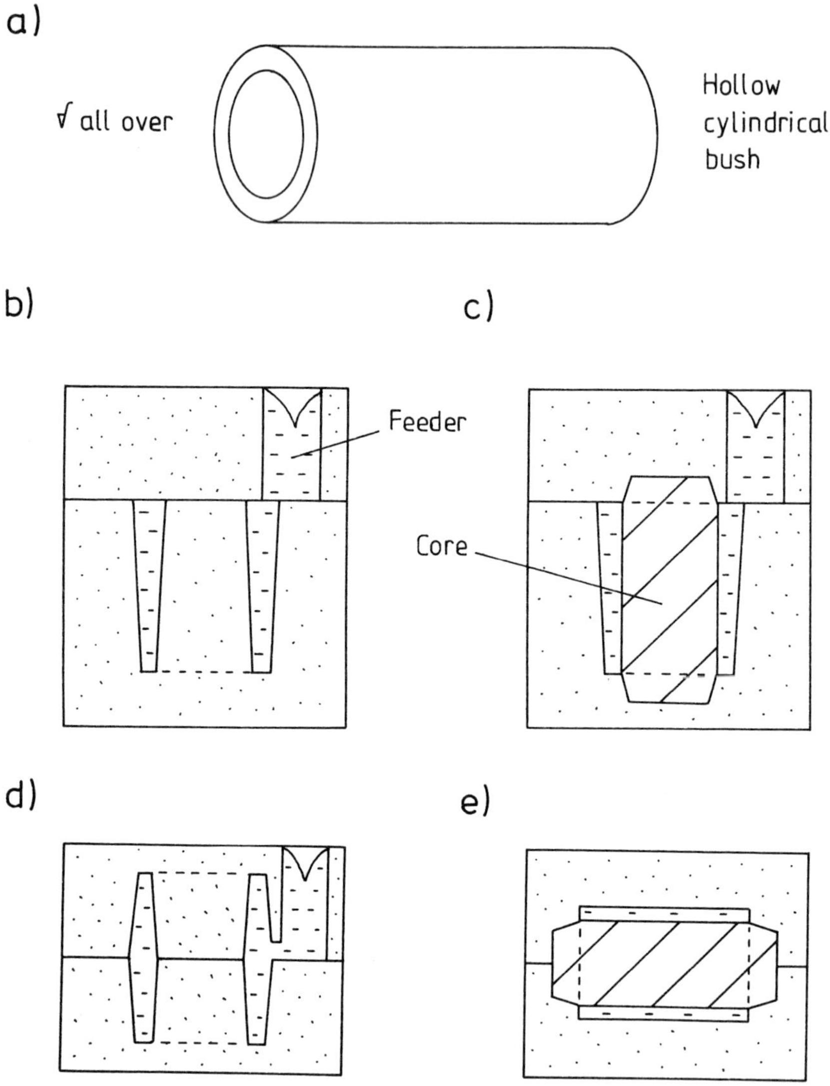


Fig. 9.1 Moulding considerations in the production of a plain cylindrical bush: a) the component shape; b) producing the casting in the drag mould without a core; c) producing the casting in the drag mould using a core; d) producing the casting with distribution equal in the cope and drag moulds; e) producing the casting in the horizontal orientation using a centre core.

The selection of the option shown in Fig. 9.1d enables a smaller mould volume to be used. The internal feature will be more readily moulded so that a core should be unnecessary in this case. Taper will still be required externally and internally on the pattern, but the volume of the casting will be smaller and the amount of material to be removed by machining will be reduced. The requirement for feeding can be met by placing a feeder at the side of the casting and feeding into the heaviest section, which is at the joint plane.

The orientation shown in option 9.1e is quite different to that shown previously and it makes the use of a core essential. In this orientation the requirement for taper is restricted to the end of the cylinder and the amount of material to be removed by machining will be significantly reduced. Whilst the cost of the casting may be increased by the cost of the core, the reduction in casting weight and machining requirement should result in a cheaper component for the customer.

Adding flanges to the bush, as shown in Fig. 9.2a, creates additional problems if a vertical orientation is chosen, because multiple joints and an increased mould volume are required, see Figs. 9.2b and 9.2c. A central core will be required and external taper will be necessary. In this orientation it would be possible to eliminate one joint from the mould if a ring core was used to form the external detail, see Fig. 9.2d. This would eliminate external taper and reduce machining costs, but at the expense of providing an additional core.

The simplest approach would be to orientate the casting horizontally, as shown in Fig. 9.2e. A central core will be required but taper will be restricted to the flanges and the mould volume would be minimised. Access for feeding to the flanges is possible at the joint plane. This represents the best practical solution.

a)

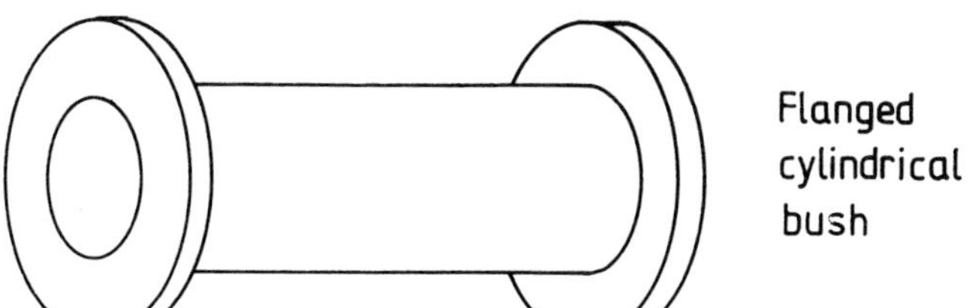


Fig. 9.2a

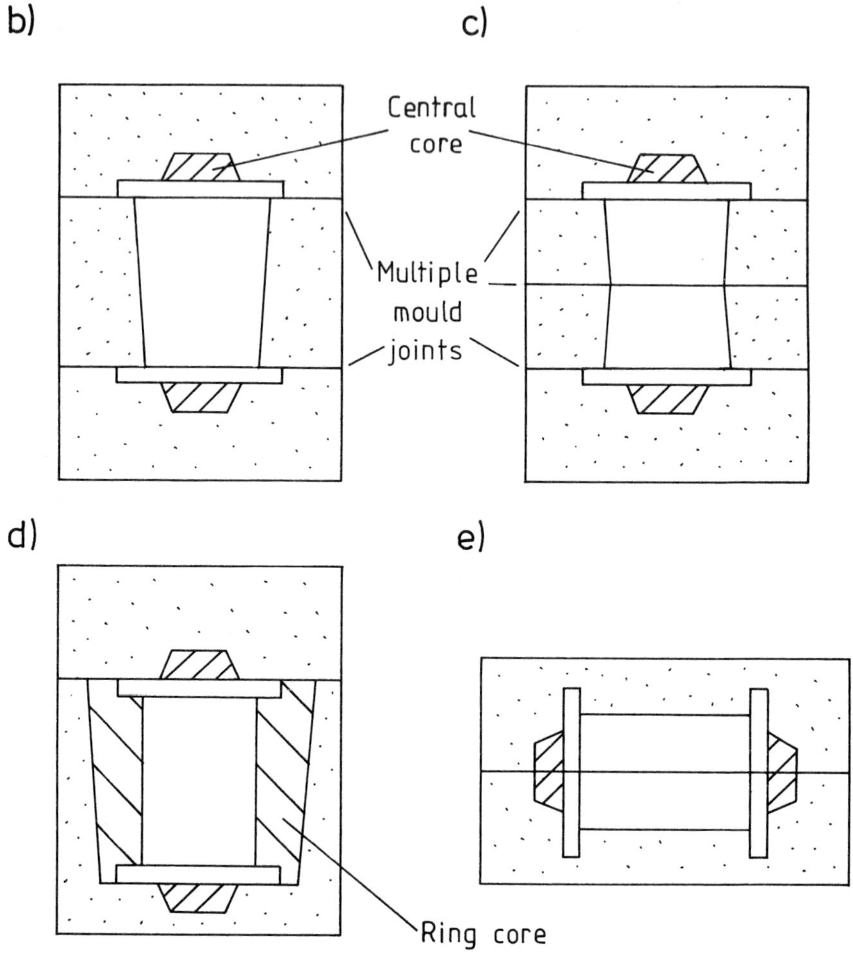


Fig. 9.2 Moulding considerations in the production of a flanged cylindrical bush: a) the component shape; b) and c) the requirement for multiple mould joints to accommodate vertical orientation of the casting; d) the need for a ring core to simplify mould production in the case of vertical orientation of the casting; e) producing the casting in the horizontal orientation using a centre core.

Figure 9.3 shows a section through the mould required to produce a flanged and ribbed bush with the ribs displaced at 60° angles. Orientation of this component in the vertical plane presents a similar situation to that described in the previous example. However, in this case, simply orientating the component in the horizontal plane provides no solution, as the angled ribs cannot be withdrawn from the mould. To overcome this problem it would be necessary to use an expensive ring core to produce the external features, see Fig. 9.3a. An alternative approach would be to redesign the component with the ribs disposed in the manner shown in Fig. 9.3b. The ribs will still provide their strengthening effect but without presenting moulding difficulties.

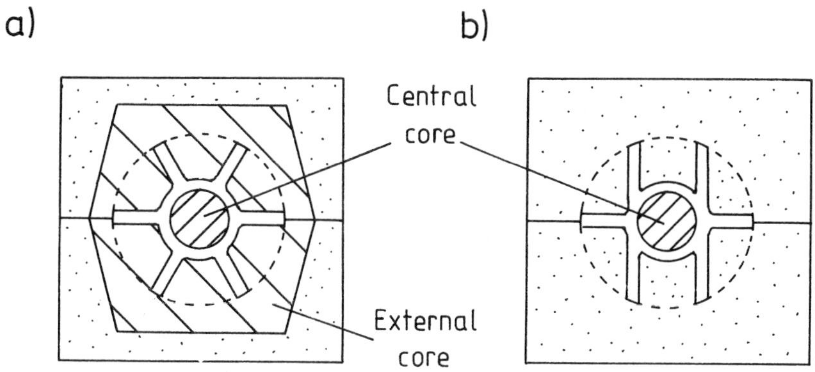


Fig. 9.3 Moulding considerations in the production of a flanged and ribbed cylindrical bush: a) ribs disposed at 60° and requiring an external core to form them; b) rib positions redesigned to permit moulding.

***Simplifying Mould Production by a Design Change.*** The simple shape shown in Fig. 9.4a cannot be moulded unless the bosses are loose and remain in the mould whilst the main body of the pattern is removed, when they can be drawn into the mould cavity for removal. Such loose pieces can be used but are not advocated. A much simpler approach is to redesign the casting, as shown in Fig. 9.4b, enabling the pattern to be removed from the mould.

***Simplifying a Mould Joint Line.*** Fig. 9.5a shows a component with rounded edges which necessitate that the joint line be positioned as shown. Minor inaccuracies and flash (metal which penetrates the joint) are likely to occur and will require removal by fettling. This is likely to impair both the dimensions and the aesthetic appearance of the component. Modification of the design to simplify the joint line, as shown in Fig. 9.5b, will improve the quality of the casting.

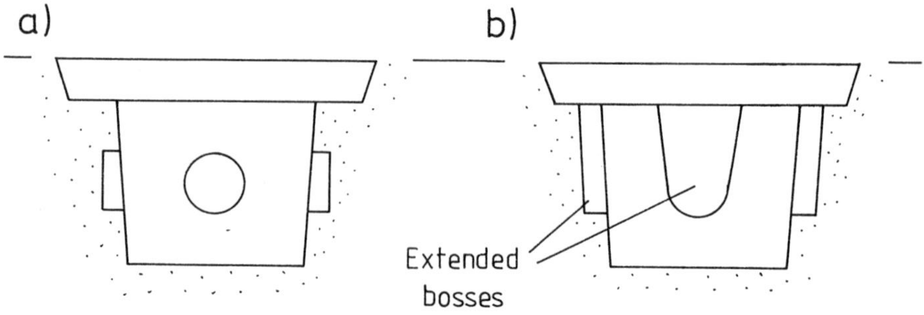


Fig. 9.4 Simplifying mould production: a) bosses which can only be moulded using loose pattern pieces; b) extended bosses to permit pattern removal from the mould.

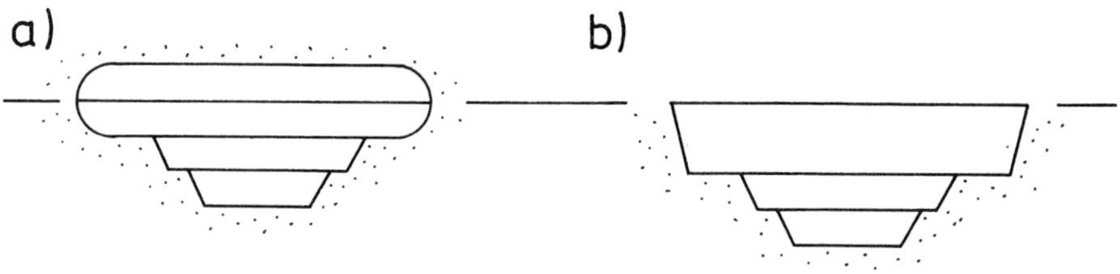


Fig. 9.5 Simplifying a mould joint line: a) a rounded edge requires a two-part pattern and may cause joint line inaccuracy; b) redesign eliminates the problem.

***Eliminating the Need for a Core.*** Cores increase production costs, increase the likelihood of casting defects and reduce the accuracy of the dimensions they form. They should not be used unnecessarily. Fig. 9.6a shows a simple shape, the design of which requires a core to produce the hollow internal feature. By redesigning the component with an external flange, as shown in Fig. 9.6b, the internal shape can be produced by moulding, thus eliminating the requirement for a core.

***Using a Core to Advantage.*** The circular component shown in Fig. 9.7a cannot be economically moulded without resort to the use of a core. Assuming that the designer insists on retaining the specific design, then the foundry would use a ring core to produce the external features whilst moulding the internal feature, see Fig. 9.7b.

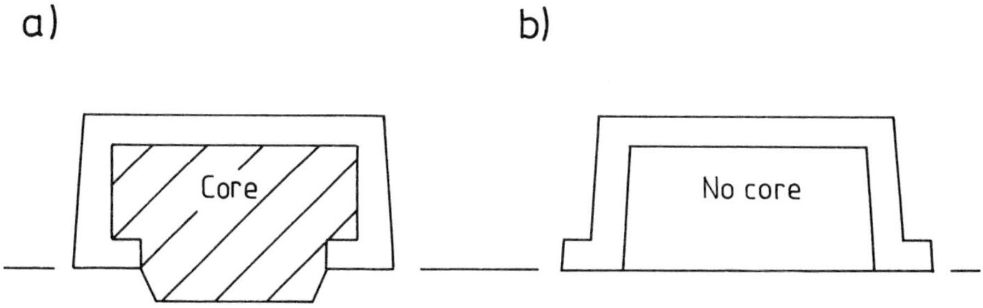


Fig. 9.6 Eliminating the need for a core: a) an internal flange requires the use of a core; b) redesigning with an external flange eliminates the need for the core.

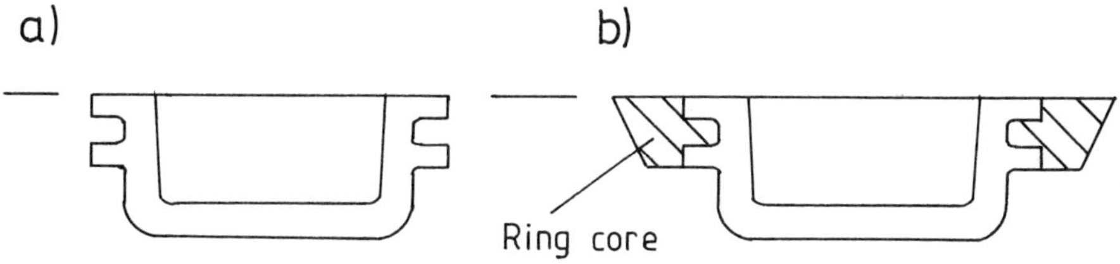


Fig. 9.7 Using a core to advantage: a) the shape cannot easily be moulded; b) the use of a ring core to produce the external features simplifies casting production.

***The Importance of Core Prints.*** Core prints are extensions of the core into a mould and provide location and support to the core, as well as an escape route for gases generated by the core binder as a result of thermal degradation. Consider the example shown in Fig. 9.8a. The core is not fixed and may be displaced laterally, thus reducing the section thickness of the casting at its closed end. The core is also likely to distort. The additional core print shown in Fig. 9.8b locates the core accurately and provides support to avoid distortion. The vent channel in the sketch shows how the gases are encouraged to escape from the core. The use of this additional core print requires a modification to the design as the casting now has two additional holes in it. This may be acceptable, depending upon the function of the component. However, if it is not but foundry practice requires the additional print, then the holes must be sealed subsequently.

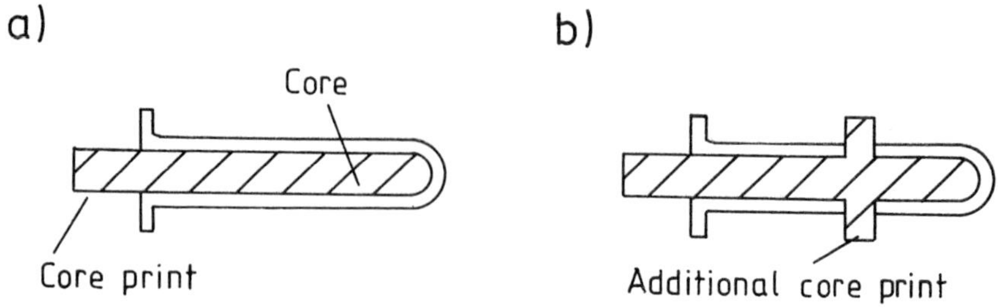


Fig. 9.8 The importance of core prints: a) the core is free to move at its remote end and may generate gas holes in the casting; b) additional core prints provide improved core rigidity and an escape route for core gases.

## DESIGN FOR SOUNDNESS

The fundamental problem of contraction was outlined in Chapter 1, when the requirement for feeding castings was described. In this chapter the requirements for gating and feeding castings will be outlined, together with the problems posed by junctions, differential contraction and the implications of as-cast structures.

### The Gating System

In the production of castings, metal is transferred from a pouring ladle or crucible to the mould cavity. The system of channels in the mould through which the metal passes into the mould cavity is termed the gating or running system and typically comprises the elements shown in Fig. 9.9.

Good gating system practice is essential if casting scrap is to be minimised and metallurgical integrity optimised. The importance of the gating system is emphasised by the list of defects which can be attributed to incorrect gating system practice, see Table 9.1.

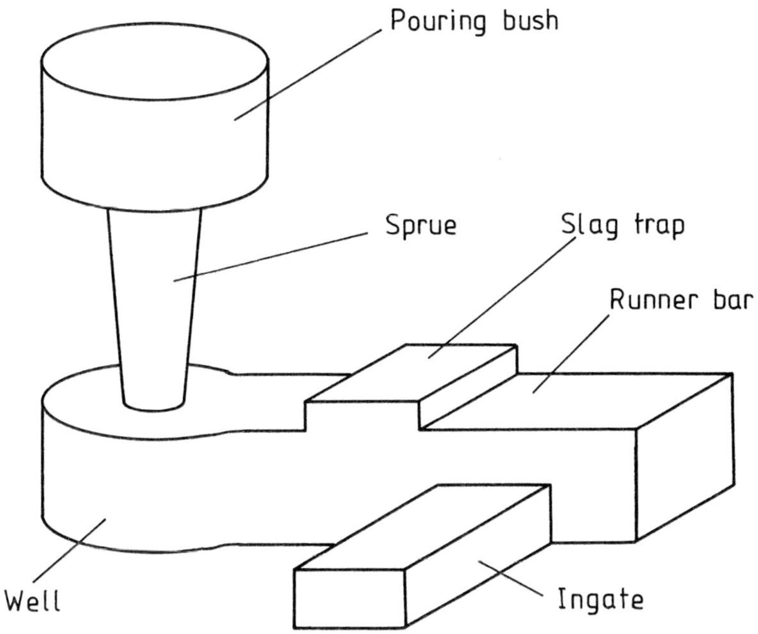


Fig. 9.9 Schematic representation of a gating system

Table 9.1 *Casting Defects which can be Attributed to Incorrect Gating System Practice*

- Sand, slag, dross inclusions.
- Rough surface finish - including metal penetration.
- Entrapped gases.
- Excessively oxidised metal.
- Localised shrinkage (pipes in ingate).
- Dispersed or micro-porosity.
- Cold laps - non-fusion of converging streams.
- Misruns.

Gating system design is based on the laws of fluid mechanics, although these assume that incompressible fluids are transported through channels with impermeable walls. Despite the fact that liquid metals are frequently transported through permeable channels and are concurrently losing heat, the laws of fluid mechanics can, with certain exceptions, be used as the basis for gating system design. The two laws which provide the basis for gating system design are: the Law of Continuity and Bernouilli's Theorem. The former states that the volume rate of flow is constant throughout the system and is a function of linear velocity and cross-sectional area of the flow passage. Thus, if the cross-sectional area decreases, the velocity must increase to maintain the flow rate, or vice versa. Bernouilli's Theorem states that the total energy of a unit weight of fluid is constant throughout a system. This energy is a function of the potential, kinetic and pressure energies, and can be summated for any point in the system. From Bernouilli's Theorem it is possible to derive an expression which relates the velocity of the metal at any point in a gating system as a function of the distance of fall from a datum point, such as the pouring bush.

The velocity of a metal stream is important for several reasons. Firstly, consider a metal stream leaving a pouring bush to travel vertically downwards through a sprue. From Bernouilli's Theorem we can predict that the velocity of the metal stream will increase as it falls. The Law of Continuity tells us that the cross-sectional area of the stream will decrease in order that the volume rate of flow remains constant. In consequence, the metal stream will leave the sides of the sprue and, if the mould is permeable and cast in air, air will be aspirated through the mould walls and be occluded into the metal stream, or cause excessive oxidation when turbulence occurs at the sprue base. Tapered sprues are advocated to overcome this problem.

High velocity streams are erosive in nature and likely to dislodge particles of mould material. They are also likely to cause penetration of the metal into the mould material when they impinge upon it. On the other hand, low velocity streams are more susceptible to aspiration and less likely to completely fill the mould cavity. Practical gating systems must take these issues into account. They are most commonly defined by their gating ratio, which is the ratio of the cross-sectional areas of the sprue, runner bar, ingate. Systems employing multiple gates present their own problems in respect of balanced flows.

***Melt Cleanliness.*** Devices for the separation of non-metallic inclusions from the metal stream are frequently incorporated in gating system designs. These inclusions may arise from the melting and transfer operations, in the form of slag or dross, or be dislodged particles of mould material. Two methods of separation are employed. The first relies upon the density difference between the inclusion and the metal. If the former is appreciably less dense than the latter it should float, and local expansion chambers can be used to trap the inclusions. Alternatively, a filter may be used to physically trap the non-metallic inclusions, and this approach is necessary when there is little density difference between the inclusions and the metal.

Non-metallic inclusions are discontinuities in the casting structure and consequently impair the mechanical properties. Inclusions may also provide suitable sites for the nucleation of shrinkage or gas porosity. As the demand for improved metallurgical integrity has increased so, too, has the recognition of the importance of metallurgical integrity and the value of filtration. Ceramic filters, cellular or foam structures, are now used extensively to reduce inclusion content, even though gravity separation might be possible.

**Principles of Feeding**

The conversion of a liquid metal from its pouring temperature to a solid casting at ambient temperature is generally accompanied by an appreciable volume contraction. This contraction occurs in three stages:

liquid state contraction
phase change contraction
solid state contraction

In the design of feeding methods for sound casting production it is the phase change contraction which is of principal concern, although the effect of superheat on liquid state contraction should not be discounted. The magnitude of solidification contraction for selected metals is shown in Table 9.2.

Table 9.2 *Solidification Shrinkage*

| **Metal** | **% Volume Contraction** |
|---|---|
| Iron | 3.5 |
| Nickel | 4.5 |
| Copper | 4.2 |
| Aluminium | 6.5 |
| Magnesium | 4.1 |
| Zinc | 4.7 |
| Lead | 3.5 |
| Tin | 2.3 |

The consequences of contraction are that the casting will be smaller than the mould cavity from which it was produced and the casting will contain shrinkage, the form of which will depend on the mode of solidification of the metal or alloy.

***Freezing Range.*** Pure metals solidify at a single temperature and the solidification process is characterised by a sharp demarcation between the solid and liquid zones. The contraction associated with solidification is readily compensated by a fall in the free liquid surface in the system. Alloys with a narrow freezing range, i.e. <20°C (70°F) between the liquidus and solidus temperatures, solidify in a similar manner. These metals and alloys are referred to as skin-forming or short freezing range (SFR). Alloys which have a temperature interval of >20°C (70°F) between the liquidus and solidus temperatures are referred to as mushy, pasty or long freezing range (LFR).

***Short Freezing Range Alloys.*** These alloys can be considered to solidify in a progressive manner from the metal-mould interface towards the centre. As solid forms it contracts, and this volumetric contraction is compensated by a fall in the level of the remaining liquid. Thus the casting demonstrates the classic primary pipe shape in its top surface, see Fig. 9.10a. The formation of the pipe is associated with a significant increase in the top surface area through which heat can be dissipated. It would not be unusual for this surface to solidify in the form of a skin whilst there is still some liquid phase present in the casting. Consequently, when this remaining liquid solidifies, there will be associated shrinkage which appears below the primary pipe at the centre line of the casting and is usually referred to as secondary piping.

To overcome the presence of this piping in the casting, a feeder head is used so that the shrinkage is contained within the head, which is then removed from the sound casting and can be recycled, see Fig. 9.10b.

***Long Freezing Range Alloys.*** Unlike the SFR alloys, these alloys show no clearly defined interface between the solid and still liquid regions. Initially, solidification will commence at the mould wall and associated contraction will cause a fall in the level of the remaining liquid. However, as the temperature of the body of the liquid falls below the alloy's liquidus temperature, solid can nucleate at points throughout the central region of the casting. As the solid grains grow to touch one another they trap liquid between them. When this liquid finally solidifies it contracts and leaves a small shrinkage cavity. The outcome of this mode of solidification is that the casting has a sound skin, shows a surface dip and contains dispersed porosity throughout its central region, see Fig. 9.11a.

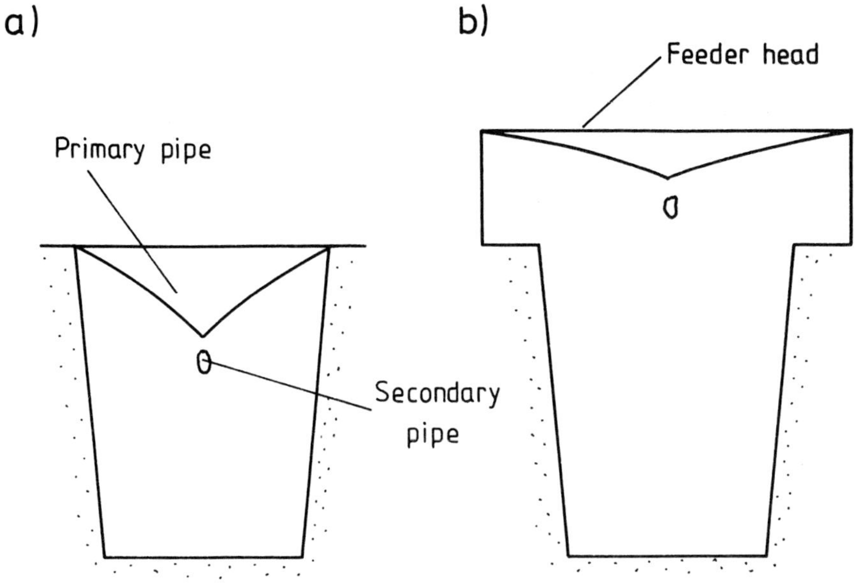


Fig. 9.10 Schematic representation of solidification in a short freezing range alloy

Producing sound castings in long freezing range alloys is especially difficult. Simply placing a feeder head in contact with the casting is not a sufficient answer. As the temperature within the head and the casting falls below the liquidus, then solid can nucleate anywhere within the still liquid region. Solidification progresses in the manner described above, so that shrinkage porosity will be evident throughout the central region of both the feeder head and casting, see Fig. 9.11b.

One way in which the problem of shrinkage in LFR alloys can be reduced is by applying the principle of directional solidification.

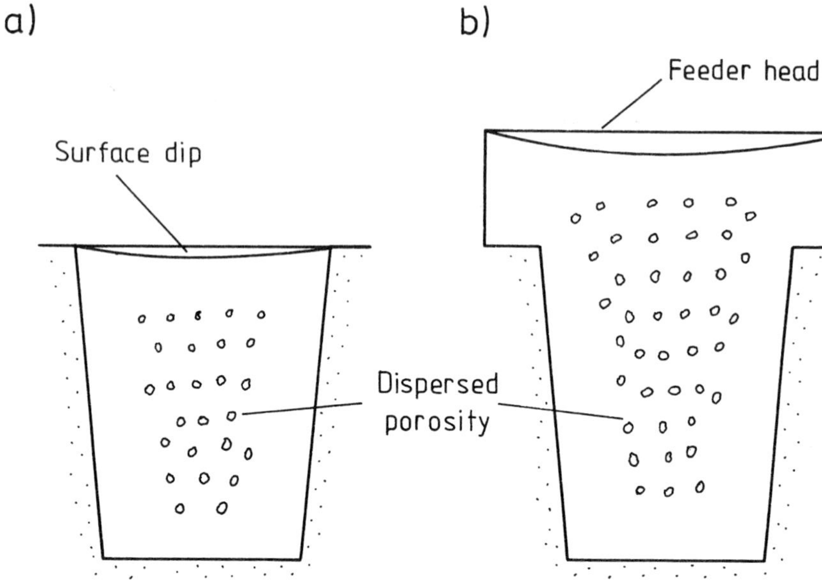


Fig. 9.11 Schematic representation of solidification in a long freezing range alloy

**Directional Solidification**

The mechanism of solidification for short freezing range metals/alloys demonstrates the basis of directional solidification. The objective of directional solidification is to promote solidification from the extremities toward the heat centre of the casting. The feeder head, or reservoir of liquid metal, is placed at the heat centre to compensate for shrinkage in the casting. It is possible to manipulate the solidification process by using a number of techniques to induce directional solidification. In practice, it is possible to make a long freezing range alloy behave as if it were a short freezing range alloy by employing these techniques.

Examples of the techniques used include:

- Tapered sections, with the widest section nearest to the feeder.
- Chills, to provide rapid heat abstraction at the extremity of the casting.
- Insulation, applied to the feeder head to extend its solidification time.

These techniques are shown schematically in Fig. 9.12.

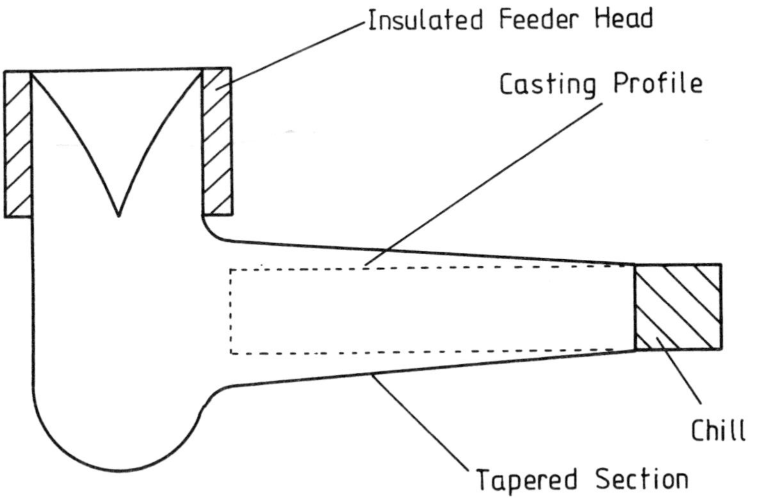


Fig. 9.12 Schematic representation of techniques used to promote directional solidification

Examples of metals and alloys which solidify according to the SFR or LFR mode are shown in Table 9.3.

Table 9.3 *Examples of Metals/Alloys which Solidify in the SFR or LFR Mode*

| **Short Freezing Range** | **Long Freezing Range** |
|---|---|
| low carbon steel (<0.2% carbon) | medium and high carbon steels (>0.2% carbon) |
| brasses | nickel-base alloys |
| aluminium bronzes | phosphor bronzes |
| high tensile brasses | gunmetals |
| aluminium | magnesium alloys |
| copper | complex aluminium alloys |

***Requirements for Feeding.*** There are four criteria which must be satisfied if the requirements for feeding are to be met, these are [3]:

heat transfer criterion
mass transfer criterion
communication criterion
pressure criterion

***Heat Transfer Criterion.*** The first requirement is that the feeder head must solidify after the casting. Whilst the laws of heat transfer can be applied to the solidification of castings, the complexity of the problem has led to the development of empirical rules. That based on the concept of modulus developed by Wlodawer [4] has found widespread favour. In principle, the concept relates solidification time to the modulus of the casting, or parts of the casting. The modulus is the volume/surface area ratio of the casting and can be determined directly from the section dimension using nomograms. To meet the heat transfer criterion the modulus of the head must clearly exceed that of the casting. Wlodawer proposes that a safety factor of 1.2 be used because the modulus of the head changes with time as it loses metal whilst feeding the casting.

***Mass Transfer Criterion.*** This criterion recognises that not all the metal in a feeder head is available to feed the casting because the head is also solidifying at a similar rate to the casting. The mass required to feed a casting is a function of the alloy's specific shrinkage and the casting's volume. However, some allowance must be made for the feeder head's inefficiency, which will be greatest for cylindrical heads moulded in the same material as the casting and least for heads which are insulated.

***Communication Criterion.*** Extended castings, such as bars or plates, present problems in feeding because, for effective feeding to occur, there must be a path through which the liquid metal can travel to reach all parts of the casting. The term feeding distance is used in this respect and experimentally derived rules [5] are available to guide the foundry engineer on this requirement. These rules provide guidance on the maximum length of bar or plate that can be produced sound as a function of bar or plate thickness. They also show how, with the judicious use of chills, feeding distance can be maximised.

***Pressure Criterion.*** This criterion recognises that a pressure difference is necessary to move liquid metal from the feeder head to and through the casting. This pressure difference will consist of the weight of the metal, i.e. the metallostatic pressure, and atmospheric pressure. Clearly, if the height of the feeder head is increased, then the metallostatic pressure will be increased. In addition to providing improved feeding, the higher pressure will suppress the formation of pores, both gas and shrinkage, by increasing the internal pressure required by a pore to grow. The use of larger (taller) feeder heads will naturally reduce the yield of the casting process.

In recognising the importance of these criteria, the designer should endeavour to ensure that a component design does not feature extended plate or bar sections, remote thick sections, or inaccessible bosses, if a sound casting is to be produced. Guidance on effective casting design may become more readily available when software is developed for solidification modelling and the computer aided design of castings.

## The Problem of Junctions

Junctions between intersecting members are a common feature in castings, they are also a potential source of shrinkage porosity unless they are designed with care. Consider the case of two members, of equal section, intersecting at right angles to form an L-junction, see Fig. 9.13a. If circles are inscribed in the members, it can be seen that the circle inscribed at the junction has the larger diameter. Wherever a junction occurs there will be a local increase in mass and the metal at this point will take longer to solidify than that in the connecting members and will, therefore, contain shrinkage porosity.

The presence of shrinkage will weaken the junction and its location may exacerbate the problem. It should not be assumed that the shrinkage will be located in the centre of the mass. Figure 9.13b shows that the rate of heat transfer is affected by the junction design. The external corner loses heat to the mould and this is dissipated rapidly through the mould material. However, the heat from the internal corner is trapped in the mould, creating a hot spot. This retards the rate of heat loss and solidification so that the final region of the junction to solidify is displaced towards the hot spot. Consequently, the shrinkage is displaced towards the hot spot and may, in fact, communicate with the internal corner if atmospheric pressure has ruptured the weak casting skin at this point.

The problem of the hot spot can be reduced by providing a radius to replace the sharp corner. However, as can be seen in Fig. 9.13c, this increases the mass of the junction and therefore the shrinkage. The problem can be overcome by designing the junction with a uniform section, as shown in Fig. 9.13d.

Junctions are frequently referred to by the letter of the alphabet that describes them. We have already considered the L-junction, the other junctions to consider are the T-, X-, V- and Y-junctions.

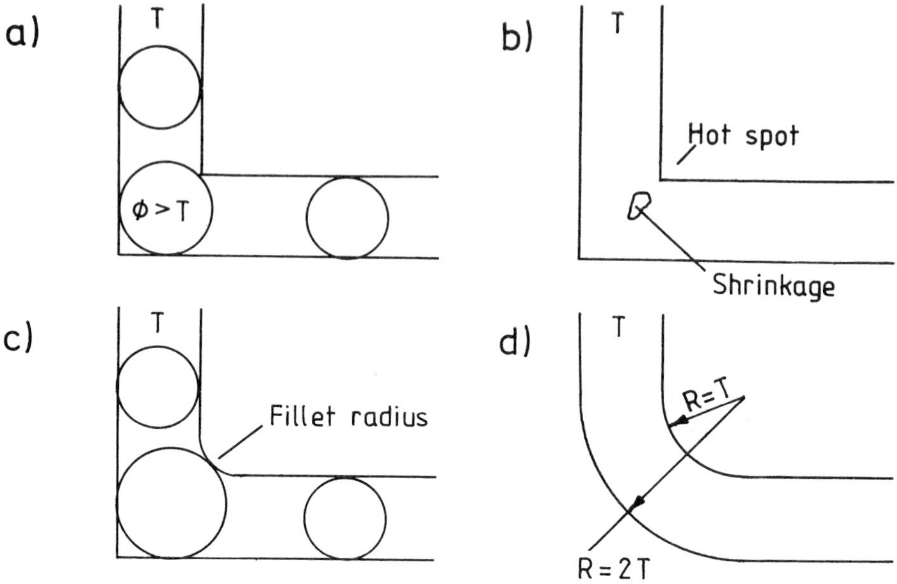


Fig. 9.13 The L-junction: a) the increase in mass associated with a junction demonstrated by the use of inscribed circles; b) the effect of a hot spot on shrinkage position; c) the increased junction mass associated with a fillet radius; d) the design recommended to eliminate junction mass related shrinkage.

***The T-Junction.*** This junction can be considered as two L-junctions back-to-back and the problems of the L-junction are magnified by the increased mass at the junction arising from the additional connecting member. Figure 9.14 summarises the approaches which can be adopted to reduce the shrinkage problem. One method involves reducing the junction mass by designing in a surface depression at the junction. This may be unacceptable on aesthetic and strength considerations. Mass may also be reduced by coring out the junction. However, the reduced section may not provide the required strength. If it is possible to place a core at a junction there will probably be access to place a feeder there and this would eliminate the shrinkage. As a final resort, the foundry could use chills to accelerate the rate of solidification at the junction so that feeding can occur through the connecting members. Whilst chills present a practical solution, they represent an additional operation which might be forgotten, a source of gas porosity from moisture which condenses on their surfaces, and a fettling problem if they fuse to the casting.

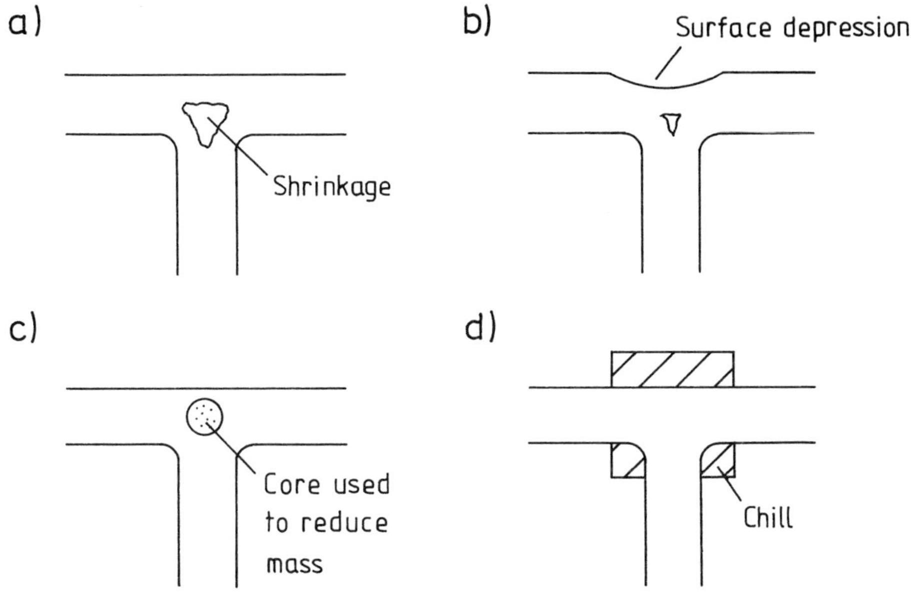


Fig. 9.14 The T-junction: a) the increased mass and resultant shrinkage associated with the junction; b) the use of a surface depression to reduce junction mass and the associated shrinkage; c) the use of a core to reduce junction mass and the associated shrinkage; d) the use of chills to accelerate the rate of solidification at the junction and permit feeding through the connecting members.

***The X-Junction.*** This junction is most often considered as a +-junction and thus extends the problem of the L- and T-junctions. The increased mass associated with this junction is considered to present an especially difficult problem to overcome. Reducing the mass by using a core is not a complete solution and most modifications to the design are effectively a substitution of the T-junction design for that of the +-junction, see Fig. 9.15.

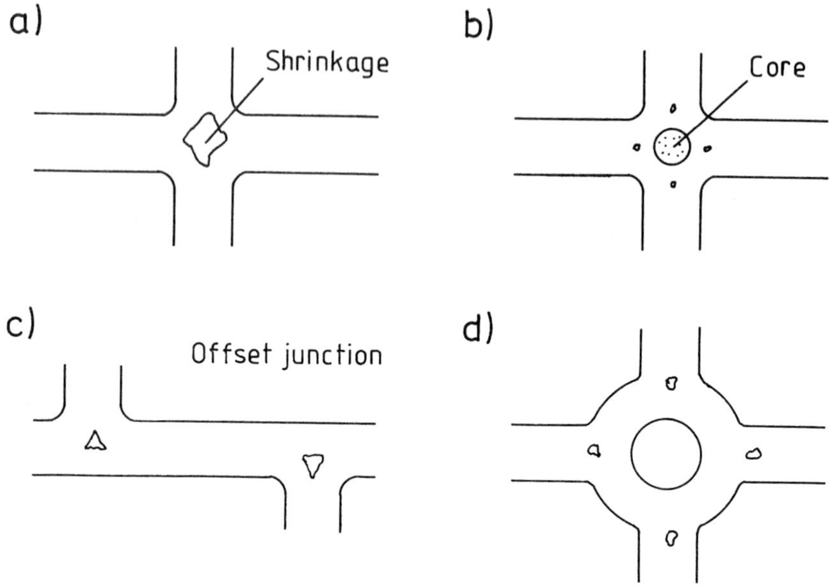


Fig. 9.15 The X-junction: a) the increased mass and resultant shrinkage associated with the junction; b) the use of a core to decrease junction mass and reduce, but not eliminate, shrinkage; c) and d) offsetting the junction simply reduces the problem to that of a T-junction.

***The V-Junction.*** Apart from the increased mass, the sharp internal angle and associated hot spot are a problem in this design. Measures which displace the connecting members and permit rounding of the junction improve matters, but to eliminate shrinkage entirely it may be necessary to reduce the section thickness at the junction, see Fig. 9.16.

***The Y-Junction.*** In this design the problems of the V-junction are exacerbated by the presence of an additional member. Again, the elimination of the sharp internal corner is essential. However, the only effective way to remove mass from the junction is by the use of a core, see Fig. 9.17.

***Junctions of Unequal Sections.*** Abrupt changes in section are unsatisfactory for similar reasons to those considered under junction design. Progressive changes are to be preferred and recommendations are provided in Fig. 9.18.

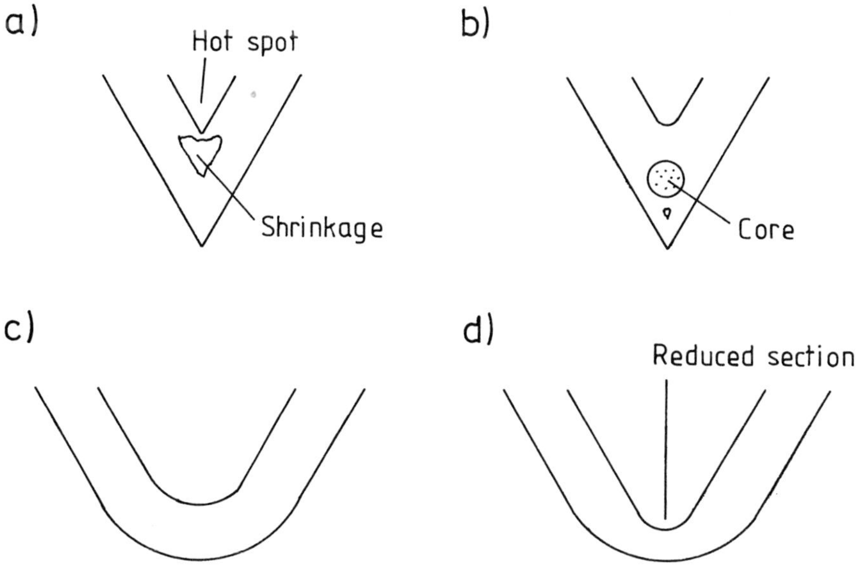


Fig. 9.16 The V-junction: a) the junction design creates a hot spot and displacement of the shrinkage from the mass centre; b) a round core will reduce the junction mass but not necessarily eliminate the shrinkage; c) lateral displacement of the joining members and a uniform section at the junction may eliminate shrinkage; d) lateral displacement of the junction arms combined with a reduced section thickness will reduce shrinkage, but the reduced section will be weaker.

## Contraction Defects

Effective junction design and feeder placement may eliminate shrinkage defects, but problems may still arise during solid state contraction. A major cause of such problems is mould or core constraint. If a casting is constrained from contracting, then stresses will be produced in the casting which may result in tears, cracks or distortion. When the problem arises during the early stage of solidification, the weak solid skin may rupture because the stress exceeds the strength of the material. This is referred to as a hot tear. It is possible for such tears to be repaired by the flow of liquid metal from within the casting, but only in the case of short freezing range alloys. Once the casting has solidified and has cooled it becomes stronger, so that a higher level of stress will be required to initiate cracking. Even if a crack does not occur, residual stresses are likely to arise. These may be relieved by a stress-relief heat treatment. If such a treatment is not provided then susceptible castings are likely to distort, particularly after machining, which removes the constraint provided by the casting skin. Examples of the problems which can arise through constraint are shown in Fig. 9.19.

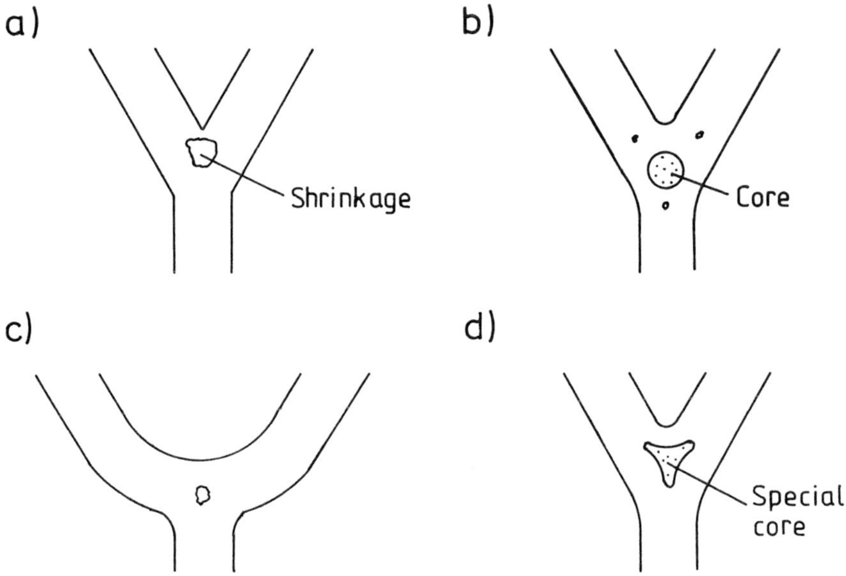


Fig. 9.17 The Y-junction: a) the junction design creates a hot spot and displacement of the shrinkage from the mass centre; b) a round core reduces the junction mass but does not entirely eliminate the shrinkage; c) lateral displacement of the upper members eliminates the hot spot but not the shrinkage; d) a special triangular shaped core can be used to reduce the junction mass and eliminate the shrinkage.

Design changes to overcome the problem of contraction are demonstrated by the spoked wheel casting shown in Fig. 9.20. The diametrically opposed spokes create the greatest stress and this problem can be reduced by changing to an odd number of spokes, thus eliminating the diametrically opposed pairs. Another approach is based on the use of S-shaped spokes, in which the stress in the spokes is relieved by lateral movement, i.e. a straightening out of the S-shape. A similar approach is demonstrated by the use of an offset boss, in this case stress is accommodated by the movement of the boss towards the centre line of the rim.

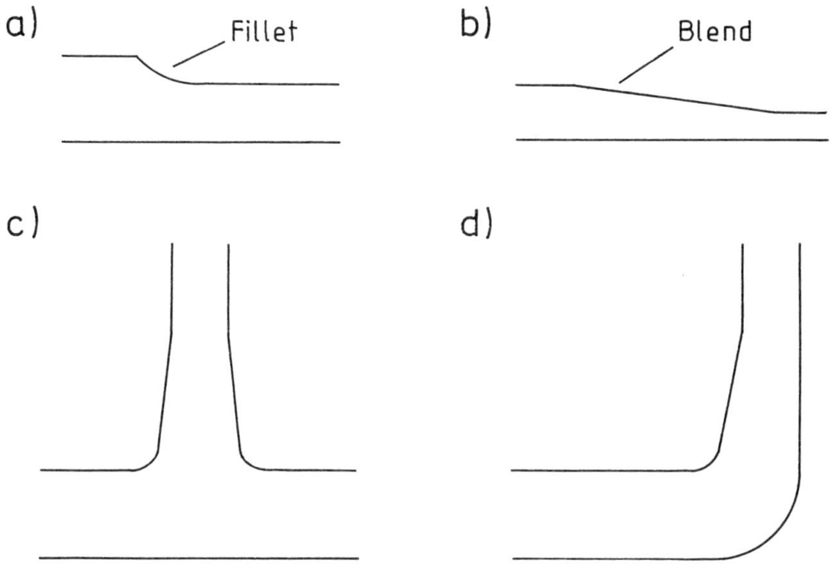


Fig. 9.18 Junctions of unequal sections: abrupt changes in section can be overcome by a) a fillet; or b) a blend in a continuous section; or by progressive changes c) at T-junctions; or d) at L-junctions.

***Cast Structure.*** Solidification exerts a fundamental influence on the properties of a casting through its effect on metallographic structure and soundness. The metallographic structure (grain size, shape and orientation, and the distribution of alloying elements) produced during solidification generally stays with the casting. The structure is influenced by such factors as alloy composition, metal treatments, mould material, and casting design. Each of these can vary, or be deliberately varied, with a consequent effect upon the structure and properties of the casting. It is beyond the scope of this book to provide a detailed account of solidification processing aspects. However, it is important to appreciate, within the context of casting design, that casting structure and properties can be varied by melt treatments, variations in moulding practice and by heat treatments.

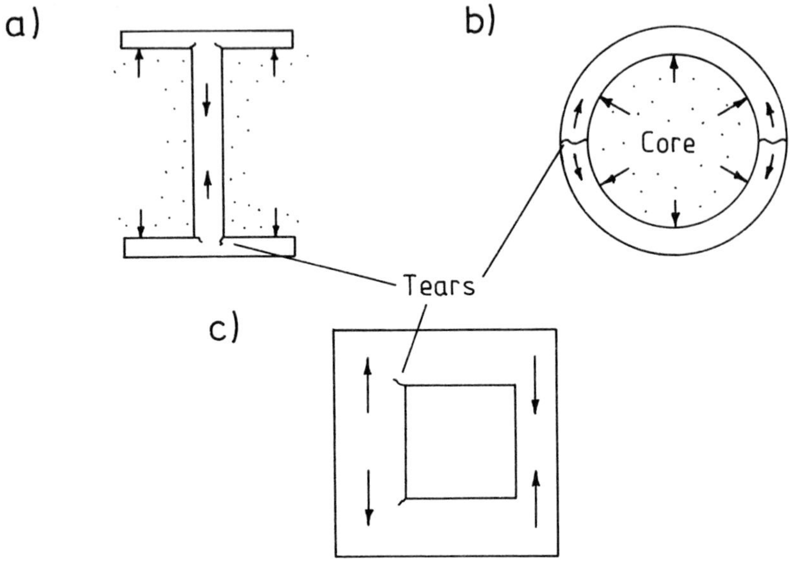


Fig. 9.19 Contraction defects: caused by a) mould constraint; b) core constraint; c) differential contraction in sections of different size.

The aspect of melt treatments will be considered in Chapter 10 and will include grain refinement treatments. Grain refinement relies upon the provision of heterogeneous nucleation sites to promote the nucleation of an increased number of smaller grains, which in turn improves mechanical properties. The mechanical properties of most cast alloys are also strongly dependent on the secondary dendrite arm spacing (DAS). As DAS decreases, so ductility and elongation increase substantially. Faster cooling rates promote reduced DAS and this provides the opportunity to manipulate cast structure without resort to melt treatments.

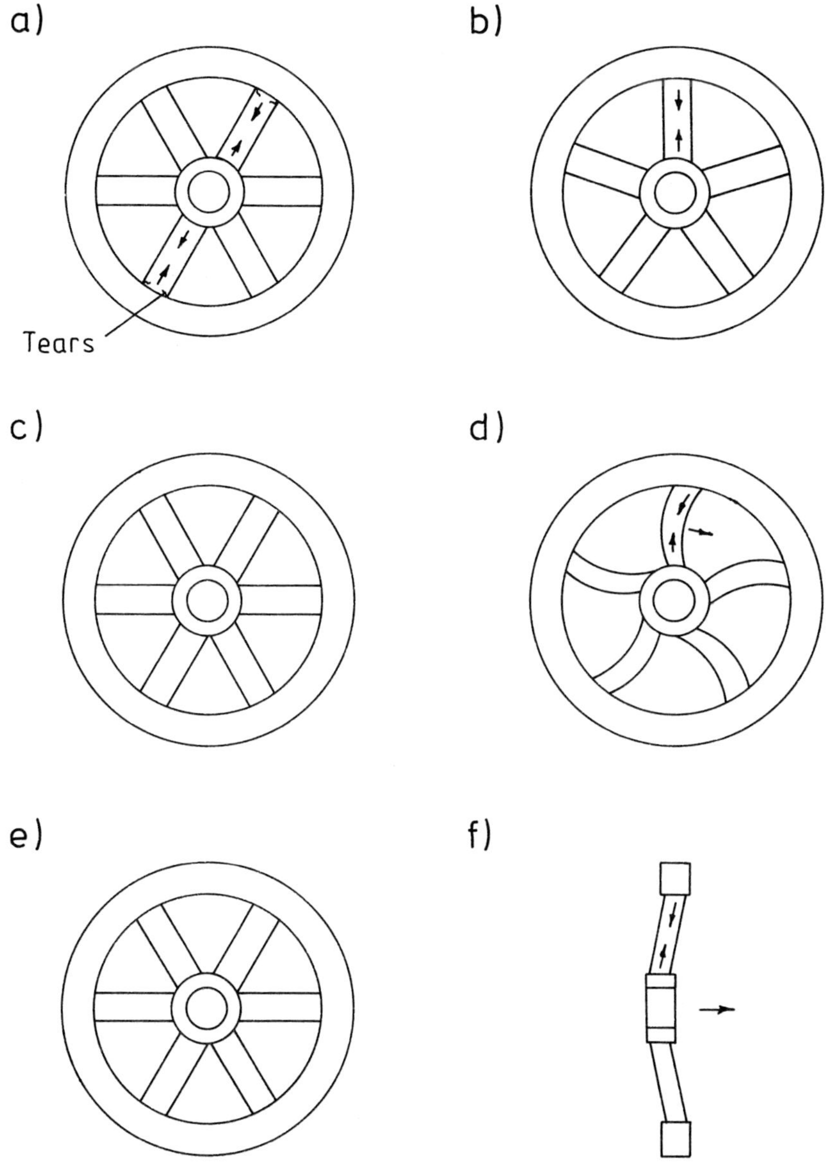


Fig. 9.20 Design changes in a spoked wheel casting to reduce contraction defects: b) elimination of diametrically opposed spokes; d) using S-shaped spokes which relieve the stress by straightening; f) offsetting the boss from the centre line and relieving the stress through the movement of the boss towards the centre line.

The effect of cooling rate on structure is a function of casting design as well as mould material. A casting design which consists of different sections may demonstrate significantly different properties in the thinnest section compared with the thickest. Thinner sections tend to be sounder and their faster cooling rate reduces DAS. Section sensitivity, that is the variation in structure and properties with variation in section, can of course be minimised by adopting a design with uniform section thickness.

**Computer Aided Design for Casting**

CAD/CAM technology has the objective of improving the speed and accuracy with which a design concept can be translated into a finished product. Although its greatest benefits may be conferred on those companies with a fully integrated manufacturing system, its implications and potential benefits will spread much further afield. The foundry industry will not remain unaffected by CAD/CAM. Foundries will find that their customers, in adopting the technology, will impose upon them the requirement to interact with their systems to ensure that they achieve the potential advantages [6].

Whilst the design of a component must remain the responsibility of the customer, the foundry can benefit its own operations as well as those of the customer by interacting at the design stage. The objective of this interaction must be to ensure that the component is designed as a casting, which implies that the requirements of the founder to meet process constraints and satisfy sound casting production can be met.

***Solidification Simulation.*** Solidification is a unique feature of the casting process, and it exerts a fundamental influence on the structure and properties of a casting. Foundrymen have long had to cope with the need to provide effective feeding systems to counteract the problem of solidification shrinkage. Rule of thumb was replaced by empirical rules, the use of which has been simplified by the adoption of microcomputer-based software programs. However, effective as these may be in providing suitable feeding methods, they make no direct contribution to the casting design process.

The macroscopic aspects of the casting process: heat transfer, thermal stress and fluid flow, are receiving considerable attention and are the most likely aspects to be incorporated in process control and process design models. Research to develop an understanding of the solidification phenomenon continues to be an important current topic and the mathematical modelling, or simulation, of solidification is a prerequisite of a dedicated casting CAD system.

In numerical computation for solidification simulation the two most important procedures are the finite difference method (FDM) and the finite element method (FEM). Originally developed for solving complex stress analysis problems, FEM has been applied to both steady state and transient heat transfer problems. The method is particularly suitable for calculating thermally induced stresses.

Although there is as yet no comprehensive package available to meet the interactive needs of the designer and the foundry engineer, it is likely that current research will provide such packages during the 1990's.

**Process Selection Summary**

When selecting a specific casting process a designer should consider the following factors [1]:

- the alloy to be cast.
- the weight of the casting.
- the size and shape of the casting.
- the minimum section thickness required.
- the surface finish required.
- the dimensional accuracy required.
- the quality required of the casting.
- the requirements for casting finishing.
- the number of castings required.
- the cost implication of the process selected.

These factors have been considered within the preceding sections and are summarised here in Table 9.4.

Table 9.4 *Summary of Process Capabilities*

| Casting Process | Alloy Range | Weight Range kg (lb) | Economical Quantity |
|---|---|---|---|
| Green Sand | Most | 0.1 to 100 (0.2 to 220) | 1 to 100,000 |
| $CO_2$/Silicate | Most | 1 to 100,000 (2.2 to 220,000) | 1 to 1,000 |
| Cold Set | Most | 1 to 100,000 (2.2 to 220,000) | 1 to 1,000 |
| Cosworth | Aluminium Alloys | 0.2 to 50 (0.44 to 110) | 1,000 + |
| Shell Mould | Most | 0.1 to 50 (0.2 to 110) | 1,000 + |
| Evaporative Pattern Casting | Most | 0.5 to 5,000 (1 to 11,000) | 5,000 + |
| Vacuum-Sealed | Most | 100 to 2,500 (220 to 5,500) | 50 + |
| Plaster Mould | Aluminium Alloys | 0.05 to 50 (0.1 to 110) | 50+ |
| Ceramic Mould | Most | 0.1 to 5,000 (0.2 to 11,000) | 50 + |
| Investment Casting | Most | 0.05 to 250 (0.1 to 550) | 1,000 + |
| CLA/CLV | Most | 0.05 to 50 (0.1 to 110) | 1,000 + |
| Replicast (CS) | Most | 1 to 100 (2 to 220) | 50 + |
| Gravity Die | Principally Non-Ferrous + Cast Iron | 0.1 to 70 (Al) (0.2 to 155) | 1,000 + |
| Low Pressure Die | Aluminium Alloys | 5 to 25 (11 to 55) | 1,000 + |
| High Pressure Die | Principally Aluminium and Zinc Alloys | 0.05 to 25 (0.1 to 55) | 5,000 + |
| Squeeze Casting | Principally Aluminium Alloy | 0.1 to 20 (0.2 to 45) | 1,000 + |

| Thin Section Capability mm (in) | Cast Hole Dimensions mm (in) | Surface Finish µm (µin) | Dimensional Accuracy mm (in) |
|---|---|---|---|
| > 5 (0.2) | > 6 (0.24) | 6.3 to 25 (250 to 1000) | - |
| > 5 (0.2) | > 6 (0.24) | 6.3 to 25 (250 to 1000) | - |
| > 5 (0.2) | > 6 (0.24) | 6.3 to 25 (250 to 1000) | - |
| > 4 (0.16) | > 6 (0.24) | 3.2 to 6.3 (125 to 250) | ± 0.15 up to 100 (± 0.006 up to 4) |
| > 2 (0.08) | > 3 (0.12) | 3.2 to 12.5 (125 to 500) | ± 0.50 up to 100 (± 0.02 up to 4) |
| > 3 (0.12) | > 3.5 (0.14) | 6.3 to 12.5 (250 to 500) | - |
| > 3 (0.12) | - | 3.2 to 12.5 (125 to 500) | - |
| > 1.5 (0.06) | > 6 (0.24) | 0.8 to 3.2 (32 to 125) | ± 0.25 up to 100 (± 0.01 up to 4) |
| > 1.5 (0.06) | - | 0.8 to 3.2 (32 to 125) | ± 0.38 up to 100 (± 0.015 up to 4) |
| > 1.5 (0.06) | > 1 (0.04) | 0.8 to 3.2 (32 to 125) | ± 0.50 up to 100 (± 0.02 up to 4) |
| > 0.75 (0.03) | - | 1.6 to 3.2 (63 to 125) | ± 0.125 up to 100 (± 0.005 up to 4) |
| - | > 9.5 (0.37) | 1.6 to 3.2 (63 to 125) | ± 0.25 up to 100 (± 0.01 up to 4) |
| > 5 (0.2) | > 6.5 (0.26) | 3.2 to 6.3 (125 to 250) | ± 0.25 up to 100 (± 0.01 up to 4) |
| > 4 (0.16) | > 6 (0.24) | 1.6 to 6.3 (63 to 250) | ± 0.55 up to 100 (± 0.022 up to 4) |
| > 0.5 (0.02) | > 1 (0.04) | 0.4 to 3.2 (16 to 125) | ± 0.15 up to 100 (± 0.006 up to 4) |
| > 6 (0.24) | - | 0.4 to 3.2 (16 to 125) | ± 0.20 up to 100 (± 0.008 up to 4) |

## References

1. Kumar, R. L. (1981). Metal casting production: considerations for design and process evaluation. *The British Foundryman,* **74**, 6, 138-141.

2. Farag, M. M. (1982). *Selection of Materials and Manufacturing Processes for Engineering Design.* Published by Prentice Hall International, London.

3. *Casting: Unit 3, Manufacture Materials Design.* The Open University Press, Milton Keynes, 1985.

4. Wlodawer, R. (1966). *Directional Solidification of Steel Castings.* Pergamon Press, Oxford.

5. Myskowski, E. T., Bishop, H. F. and Pellini, W. S. (1952). Application of chills to increase the feeding range of risers. *Trans AFS*, **60**, 389-400.

6. Clegg, A. J. (1986). Foundry CAD/CAM: an overview. *The British Foundryman,* **79**, 8, 397-400.

## CHAPTER 10

# Metal Melting and Treatment

## INTRODUCTION

The metallurgical integrity of a casting is a function of casting design and the methods used to gate and feed the casting. These issues were considered in the previous chapter. However, metallurgical integrity is also a function of the metal quality. The metal melting stage exerts a fundamental influence on the quality of the metal. The chemical composition of an alloy is specified to a close tolerance and the first objective of melting is the attainment of the specified alloy composition. Adjustments to the composition may be necessary and can include refinement, to reduce specific element levels, and/or additions to increase levels. Adjustments of this nature require the availability of analytical facilities to provide rapid data on the chemical analysis of the melt.

Whilst they are molten, certain alloys are susceptible to gas absorption and/or the loss of volatile or reactive elements. Special treatments, such as degassing or vacuum melting and casting, may be required to overcome such problems. Careful control of the metal temperature is essential as these problems are exacerbated by higher temperatures. Furthermore, high metal temperatures will accentuate shrinkage problems and increase the likelihood of metal mould reactions. It is, of course, necessary to superheat the metal to provide sufficient fluidity for casting.

The opportunity also exists at the metal melting stage to treat the metal in order to modify the structure and, consequently, the mechanical properties of the casting. These treatments are usually carried out at the point of dispensation of the metal from the furnace and include deoxidation, grain refinement, modification and inoculation. It is the purpose of this chapter to outline the procedures for metal melting and treatment with reference to selected alloy systems.

## METAL MELTING

### Furnace Classification

Furnaces may be classified according to their purpose, mode of heating, shape, or method of charging [1]. In this section only those furnaces used for melting within the foundry industry are considered, with classification being primarily according to shape.

***Crucible Furnaces.*** Furnaces of this type are widely used for melting small batches of ferrous and non-ferrous metals and alloys. The charge is melted in a refractory container, the crucible, generally by an external heat source. Heat may be provided by solid, liquid or gaseous fuels, or by electricity. The crucible design is such that the minimum surface area of the melt is exposed to the atmosphere, thus minimising oxidation and gas absorption. Small units may consist of a fixed furnacc body from which the crucible containing the melt is removed. Larger units may consist of a tilting furnace body in which the crucible is retained. In the latter case the crucible life is extended because there is less thermal and mechanical stress. However, there may be a need for a separate ladle to transfer the metal from the furnace to the mould. This may require additional melt superheat, to combat the loss of temperature during transfer, and will introduce turbulence and the danger of oxidation and gas absorption in susceptible alloys. Heat transfer in the crucible furnace is essentially by conduction. Examples of crucible furnace types are shown in Fig. 10.1.

***Hearth Furnaces.*** In this type of furnace the charge usually resides on the floor of the furnace and is heated from above by the energy source. As the mechanism of heat transfer is primarily by radiation, a relatively large melt surface area is required. This may be a disadvantage because it increases the opportunity for oxidation and gas absorption, but is advantageous if refining of the melt is necessary because it maximises the contact area between the melt and the refining slag cover. One example of this type of furnace is the direct arc furnace, in which the current from the electrodes passes through the slag to the metallic charge, forming three large arcs of plasma with temperatures of several thousand degrees centigrade. A second type is the single phase, indirect arc furnace, in which the arc is struck between two carbon electrodes with the energy being radiated to the charge directly and by reflection from the furnace roof. Rotating or rocking motions are often employed to improve heat transfer to the melt and prolong furnace refractory life. Schematic diagrams of hearth furnaces are shown in Fig. 10.2.

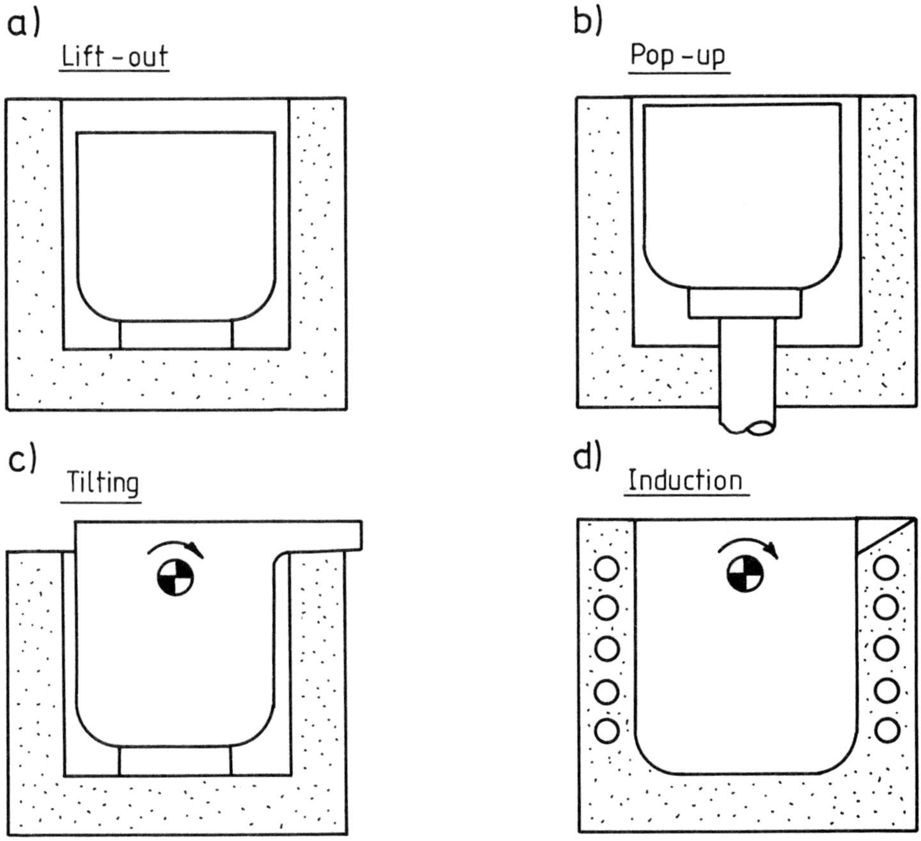


Fig. 10.1 Schematic representation of crucible furnaces

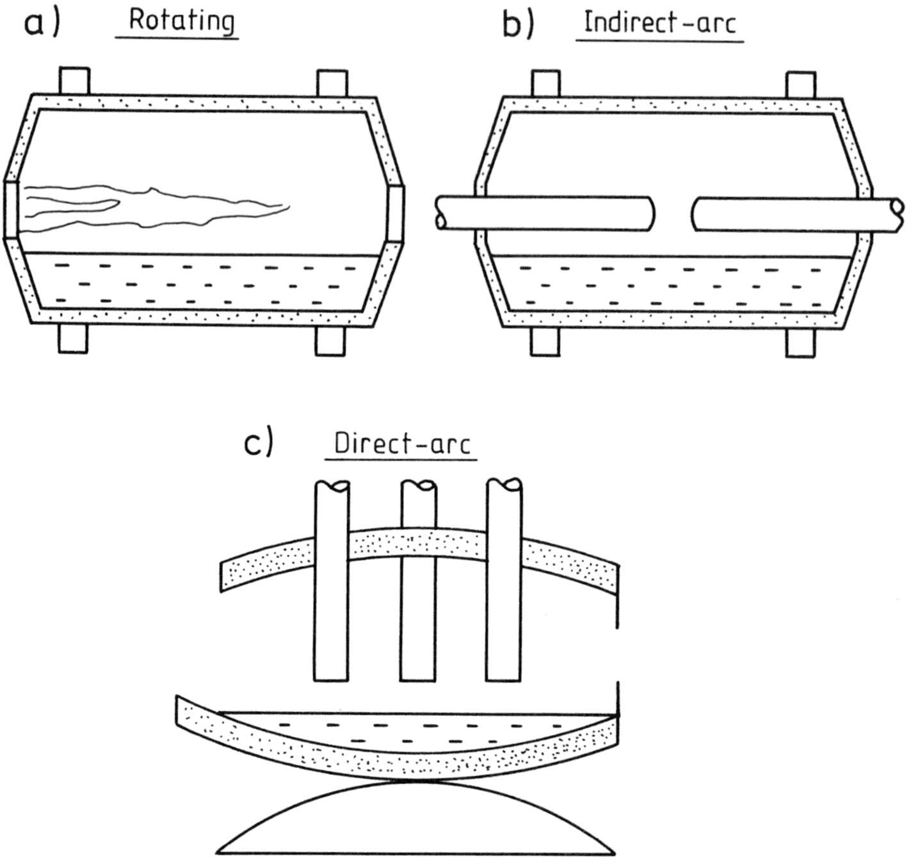


Fig. 10.2 Schematic representation of hearth furnaces

***Shaft Furnaces.*** The principal example of this type of furnace is the cupola, which is still used extensively in the production of cast iron. Similar in principle to the blast furnace, the cupola is heated by the combustion of coke, with the ascending heat transferred to the descending charge. Superheating of the melt occurs as the metal trickles through the incandescent coke to collect in the well, from which it is tapped. A schematic diagram of this furnace is shown in Fig. 10.3.

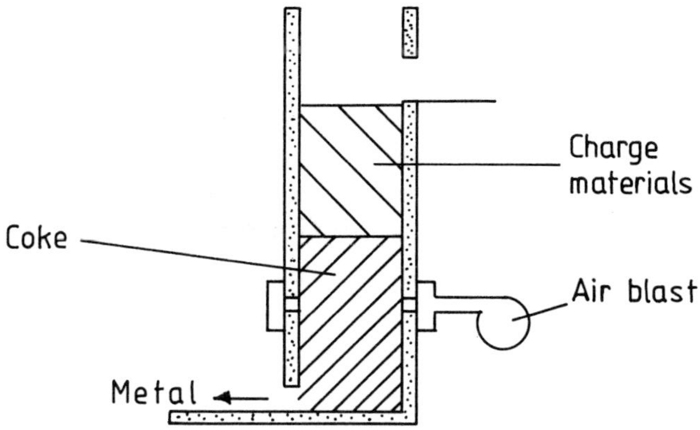


Fig. 10.3 Schematic representation of a cupola shaft furnace

**Induction Melting**

The principle of induction heating can be applied to the melting of both ferrous and non-ferrous metals and alloys. This method of melting has grown in popularity because of the excellent metallurgical control it provides, combined with improved productivity and relatively pollution-free operation. The system is designed so that the magnetic field produced by the coil interacts with the charge, which is held in some form of container [2]. Although there are two types of induction melting furnace, coreless and channel, it is the former which is most widely used for metal melting.

***The Coreless Induction Furnace.*** The basic design of the coreless induction furnace is shown schematically in Fig. 10.1d. Furnaces are available with a capacity of a few kilogrammes (pounds) to more than 100 tonnes (tons) and are operated at frequencies between mains frequency and 10 kHz. The coils are constructed from hollow copper tubing, to permit water cooling, and the loops are insulated from each other with tape. The furnace unit is constructed as a rigid structure to withstand both the electrical and mechanical forces which arise during operation. For maximum efficiency the gap between the coil and the metal charge should be as small as possible. To achieve this, rammed monolithic linings or thin preformed crucibles are used. The coil and lining are mounted on a refractory base and supported using a steel structure. In order to minimise the induced current heating of the steel structure, packs of low-loss laminations are arranged around the structure to divert the flux.

When the furnace has been loaded the power is switched on and current is induced in the charge. The nature, shape and size of the charge material will determine the frequency required. It is difficult to start furnaces operating at mains frequency from cold and it is usually necessary to leave a molten metal 'heel', equivalent to 15 to 20% of the furnace volume, in the furnace. This is clearly only acceptable when consecutive melts are of similar composition. Furnaces can be started from cold at medium frequencies, up to 1 kHz, providing that finely divided charge material is not used.

Once melting is well under way the currents in the melt give rise to mechanical forces which produce stirring, see Fig. 10.4. These may be beneficial to melt homogeneity, but can impair refractory life and expose the melt surface to atmosphere. The stirring action is proportional to the power input and inversely proportional to the frequency. The actual shape of the furnace, in terms of length to diameter ratio, is important both from the view of stirring action and power transfer. A ratio of 3:2 would be typical. The number of turns required on a coil varies inversely with frequency. However, because of insulation and cooling requirements, the coil design is limited to around 80 Ampere-turns/mm (2000 Ampere-turns/in) height of furnace.

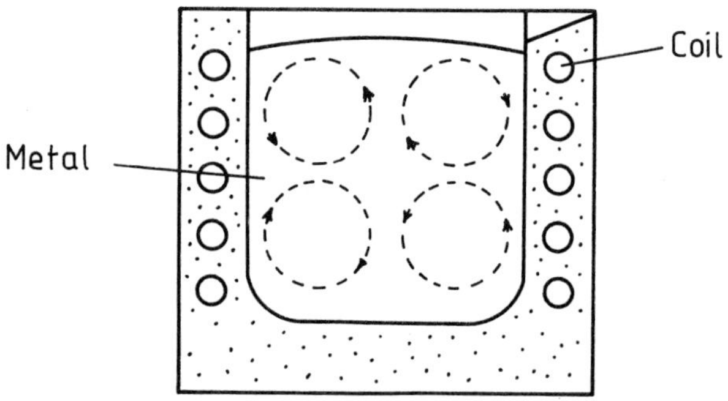


Fig. 10.4 Stirring effects associated with induction melting

Coreless induction furnaces have the advantages of high melting rates, reduced metal loss, freedom from contamination, low running and maintenance costs, and high quality output. Induction melting may also be conducted under vacuum conditions, which further improves melt quality.

## GASES IN METALS

Castings may contain gases in one of several forms, which include: gases in chemical compound form, gases in solution, or gas cavities. The presence of gases may arise during melting when gases may dissolve in the melt or react with it to form oxides or other compounds. Alternatively, gases may be entrapped during pouring or be generated by reaction between the metal and the mould material.

### Gas Absorption During Melting

The potential for gas contamination of the metal is very high at the melting stage. Reaction and solution rates usually increase with an increase in temperature and the solubility of gases is generally greater at higher temperatures. Whilst accepting that these problems are to some extent inherent at the melting stage, every precaution should be taken to minimise gas absorption at this stage.

***Melting Precautions.*** The charge materials used in melt preparation can be a source of gas contamination. These materials may already contain dissolved gases from a previous melting stage or surface contamination in the form of oxides, corrosion products, paint, lubricants, or water. Careful selection of material, proper storage and preheating, can therefore reduce the problem of gas contamination. Additions to the melt in the form of refining or fluxing agents and alloys may also be potential sources of gas contamination and must be treated with similar care. Refractories used for furnace construction, crucibles, ladles, launders, and coating furnace tools, must be sufficiently preheated to remove both free and combined moisture. The furnace atmosphere will always be a source of contamination when metals are melted in air. However, the problem will be greatest when fuel-fired furnaces are used and the products of combustion, including water vapour, can contaminate the melt.

### Degassing Procedures

Although precautions at the melting stage may minimise the problem, it is difficult to produce a melt which is free of gas contamination under normal foundry air melting conditions. Degassing procedures are conducted to remove those gases which are soluble in the liquid metal. Two problems can arise through the presence of such gases. Firstly, their solubility in the liquid metal is high but generally decreases with a decrease in temperature and is dramatically lower in the solid metal. Consequently, if the gas is not removed, its precipitation upon solidification will result in the formation of gas holes. Hydrogen is a prime example of a gas which behaves in this manner and is a potential problem in several alloy systems.

Oxygen does not usually precipitate in this manner as it is more likely to form stable metal oxides. However, when it exists in the soluble form it may react to form a compound gas. In steel it may react with carbon to produce carbon monoxide and in copper-base alloys it may react with hydrogen to produce steam.

A number of approaches are used in practice to degas melts and include those described in the following sections.

***Gas Flushing or Scavenging.*** When a melt is held under equilibrium conditions with its atmosphere, the amount of gas dissolved within the melt will be a function of the partial pressure of that gas. Simply stated, the greater the concentration of the gas in the atmosphere above the melt, the higher the contamination of the melt by the gas. If the atmosphere above the melt were to be replaced by one which had a low concentration of the contaminant gas, then the gas would diffuse from the melt to the atmosphere until an equilibrium condition was established. This provides a mechanism for degassing and its efficiency can be dramatically improved by bubbling the atmosphere through the melt.

***Oxidation/Deoxidation Practice.*** In a number of alloy systems the presence of oxygen in a melt reduces the melt's capacity for absorbing hydrogen. This feature can be deliberately exploited by melting under oxidising conditions to prevent hydrogen dissolution. However, as the oxides may be harmful to the properties of the alloy, an effective method of deoxidation must be practised prior to casting.

***Deoxidation.*** Although deoxidation is often considered to be a melt treatment in its own right, it is in effect a degassing treatment. However, whereas the gas flushing methods seek to remove the contaminant in gaseous form, deoxidation relies on the tendency of oxygen to form stable oxides which are then removed to the slag.

### Vacuum Melting and Casting

The problems of melt contamination and composition control can be largely overcome by melting and casting under vacuum conditions. Elimination of an air atmosphere during melting prevents oxidation and gas absorption. In addition, gases already in solution may be removed because the reduced pressure above the melt aids in their nucleation and their diffusion is promoted by the low partial pressure of the gas above the melt. Vacuum induction melting and casting is considered to be an essential requirement for processing certain heat resisting alloys. The practice is not without certain drawbacks. For example, prolonged exposure to vacuum may result in the loss of certain volatile elements and melts are more likely to be contaminated by reaction with the furnace refractories. Nevertheless, the advantages outweigh the disadvantages.

The capital cost of vacuum melting and casting equipment is very high. However, the advantages of vacuum degassing can be obtained without the need for an expensive, fully integrated melting and casting unit. Air melted material can be subjected to a vacuum degassing treatment prior to casting. Industrially established processes include: static bath treatment, induction degassing, fractional degassing, and stream droplet degassing.

## MELT TREATMENTS

The practices of degassing and deoxidation are treatments which improve the cleanliness and quality of the melt. However, there are a number of treatments which can be practised that will modify the metallurgical structure of the castings and consequently their mechanical properties. To understand how these treatments work it is necessary to consider the mechanisms of nucleation and growth of solid from liquid metals. However, the subject is beyond the scope of this book and the reader is referred to the work of Flemings [3] for a detailed consideration of solidification processing. For our purposes we need only consider that solidification requires the successive stages of nucleation and growth. In commercial alloys nucleation occurs heterogeneously, this implies that there are sufficient impurities present within a melt to provide sites upon which solid can form. Clearly, the more sites at which solid can nucleate the more grains will be produced and the smaller they will be. This provides a potential mechanism for changing the structure of a casting. The deliberate addition of suitable nuclei to a melt will increase the number of solidification sites and thereby decrease the size of the grains, this is the principle of grain refinement.

In this section, selected alloy systems will be considered and details of the melt treatments practised will be provided as a means of demonstrating how melt quality or metallurgical integrity can be improved.

## Non-Ferrous Alloys

***Aluminium-Base Alloys.*** These alloys are highly susceptible to oxidation and hydrogen absorption when in the molten state. This leads to a degradation of the mechanical properties of the alloys as a result of oxide inclusions and/or gas porosity. In each case the problem is exacerbated by high temperatures, which emphasises the requirement for careful control of melt temperature. Aluminium oxide in the form of films or particles impairs the fluidity of the metal and reduces the flow of feed metal. The inclusions also provide nucleation sites for the formation of both gas and shrinkage porosity. Cover fluxes provide a barrier to the reaction between the melt and the atmosphere and can absorb any oxides present at the surface of the melt. Suspended oxides and other non-metallic inclusions can be removed by using chemically active fluxes which de-wet the particles and remove them physically to the surface of the melt for removal by skimming. Porous foam ceramic filters are being used increasingly to remove suspended non-metallic inclusions from aluminium alloys and provide a demonstrable improvement in mechanical properties. However, their use should not be seen as a substitute for careful melting practice.

Hydrogen is the only gas that has appreciable solubility in aluminium alloys and its solubility versus temperature relationship is shown schematically in Fig. 10.5. Every effort should be made to minimise the presence of hydrogen in the melt by exercising the precautions referred to in the section on gases in metals. However, to reduce it to an acceptable level, a degassing treatment will be necessary. Gas fluxing techniques using chemically pure nitrogen, argon, chlorine or freon are effective, as is the use of compounds, such as hexachlorethane, which dissociate in the melt to generate a fluxing gas.

A fine grain structure improves mechanical properties, response to heat treatment, pressure tightness, hot tear resistance, and feeding characteristics. Certain aluminium alloys can be grain refined by the addition of alloys of titanium, or of titanium and boron, in aluminium. The formation of titanium aluminide, titanium boride and titanium carbide (carbon may be present as an impurity) provides the nuclei for effective grain refinement. However, it is possible that such nuclei may also promote porosity through the nucleation of shrinkage and gas porosity.

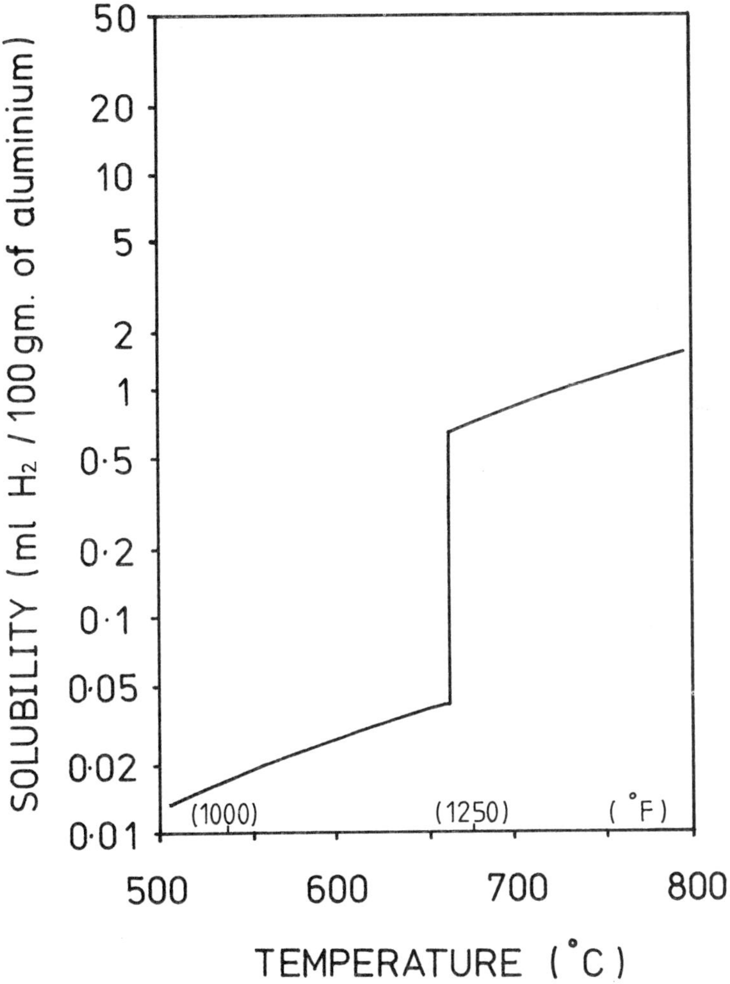


Fig. 10.5 The solubility of hydrogen in aluminium as a function of temperature

Hypoeutectic aluminium-silicon alloys consist, in the solid state, of primary α (aluminium with a small amount of silicon dissolved in it) in a matrix of α + β (β is silicon with a small amount of aluminium dissolved in it). The eutectic β will be large in size and coarse in shape when the alloy is slow cooled and this will impair strength and ductility. Faster rates of cooling, as encountered in die casting, refine the size and shape of the β phase and this improves strength and ductility. A similar effect can be obtained by a chemical addition of an element known as a modifier. Sodium is the preferred modifying agent and may be added in elemental form (vacuum packed) or as a salt in the cover flux. Very low sodium concentrations provide effective modification, although its effect fades with time and is lost on remelting.

Certain aluminium alloys, particularly those containing about 4% of copper, are amenable to heat treatment. Although clearly not a melt treatment, such treatments provide a valuable method of optimising the properties of cast alloys. In the case of such aluminium-copper alloys the heat treatment procedure involves solution treatment followed by age or precipitation hardening [4].

***Copper-Base Alloys.*** A wide range of copper alloys is produced by casting and includes: brasses; aluminium, manganese, nickel, phosphor, silicon, and tin bronzes; copper-nickel alloys; copper-chromium alloys; beryllium-copper alloys; etc. The melt treatment procedure will depend on alloy composition, but most copper alloys are susceptible to hydrogen absorption. For many of these alloys the approach adopted for minimising hydrogen contamination is that of oxidation/deoxidation. The metal is melted under oxidising conditions and subsequently deoxidised before pouring.

In general these alloys are not grain refined, although certain alloys will respond favourably to the practice. Similarly, certain alloys will respond to heat treatment. The beryllium-copper and chromium-copper alloys respond to solution treating and ageing, and the aluminium bronzes (with >10% aluminium) will respond to a quench and temper treatment.

***Magnesium-Base Alloys.*** These alloys are highly susceptible to oxidation during melting so that special precautions must be taken to prevent this, including the careful selection of refractories. Melting can be conducted under flux covers, but cleaner melts are produced when a protective gas atmosphere is used. The presence of carbon dioxide and sulphur hexafluoride is effective in reducing the extent of oxidation. The magnesium-aluminium alloys can be simultaneously degassed and grain refined by plunging hexachlorethane tablets into the melt. The presence of zirconium provides effective grain refinement for the magnesium-zirconium alloys.

The surface of the molten metal must be protected during pouring, either by the protective gases described or by dusting with sulphur and boric acid. These alloys are highly susceptible to metal-mould reactions when the mould contains water. In such reactions the water dissociates to produce hydrogen, which dissolves in the metal, and a metal oxide. It is necessary to include inhibitors such as boric acid, potassium fluoborate or ammonium fluosilicate in the sand formulation to overcome this problem.

Certain alloys are amenable to heat treatment by solution treatment and age hardening.

***Titanium-Base Alloys.*** These alloys represent a specialised and small area of casting production, but one which is growing. The molten alloys have an especially high affinity for oxygen, nitrogen and hydrogen. Melting is conducted under vacuum and special arrangements are required to ensure that titanium does not react with the oxygen in refractory materials. The principal method of melting titanium alloys is the vacuum consumable electrode method. In this method, a consumable titanium alloy electrode is lowered into a water-cooled copper crucible. A titanium-alloy skull or lining solidifies against the crucible so that no direct contact occurs between the molten alloy and the crucible. When sufficient alloy has melted it is quickly poured into the mould. As controlled superheating presents problems in this situation, melt treatments are not practised. Post casting treatments include hot isostatic pressing, to eliminate porosity, and heat treatment procedures based on solution treatment and ageing have been developed.

***Zinc-Base Alloys.*** These alloys, which are mainly processed by pressure die casting, are relatively straightforward to melt and cast. A high proportion of remelt material can be tolerated and, provided that undue agitation of the melt is avoided and the temperature is controlled, oxidation and the loss of aluminium or magnesium will be minimised. Zinc-alloy melts are not normally degassed, deoxidised, grain refined, or modified in any way. Certain alloys age and shrink with time so that castings may be subjected to a low temperature stabilisation heat treatment to accelerate the ageing and prevent its occurrence during service life.

## Ferrous Alloys

This group contains the cast irons and steels, the former are considered first as they are somewhat easier to process.

***Grey Cast Iron.*** Cast iron is a generic term for a whole family of irons, of which grey cast iron is the progenitor. These alloys, in which the principal alloying elements are carbon and silicon, can be melted by cupola or induction melting. Under normal circumstances they are not susceptible to oxidation or gas absorption. However, hydrogen porosity may arise in the presence of trace aluminium, and high nitrogen levels, generally associated with urea binders in the sand, will also produce gas defects. The characteristics of grey cast iron are a function of the amount and form of the carbon, which is normally present as graphite flakes. Thin sections are prone to solidify white because the faster cooling rate promotes the formation of iron carbide rather than graphite. To promote the formation of graphite, the iron may be inoculated at the ladle filling stage by the addition of certain silicon-containing alloys.

Grey cast iron does not respond to heat treatment, although stress relief annealing can be practised.

***Spheroidal Graphite Cast Iron.*** Spheroidal graphite (SG) or nodular cast iron are the terms used to denote that the graphite appears in the form of spheroids or nodules rather than flakes. The absence of crack propagating flakes enables a high strength iron possessing ductility to be produced. The production of SG iron requires careful attention to melt composition and processing. To achieve the best combination of strength, ductility and toughness, raw materials must be chosen that are low in those trace elements which promote a pearlitic matrix. The spheroidal form of graphite is usually achieved by adding magnesium to the melt. As this element is highly reactive at the processing temperatures involved and combines readily with oxygen and sulphur, oxidation must be minimised and the metal desulphurised. Several methods of treating melts with magnesium are practised. The effect of magnesium fades with time and the presence of the element in the iron creates a dross forming tendency which is exacerbated by poor gating practice.

The pearlitic grades of SG iron are amenable to hardening and tempering heat treatment practice.

***Malleable Cast Iron.*** These irons were the original 'ductile irons' and their popularity has declined as that of SG irons has increased. The composition of these irons is deliberately chosen so that they will solidify white. A special heat treatment procedure is required to convert the carbide to graphite, which forms as aggregates in either a ferritic or pearlitic matrix. The pearlitic grades are amenable to hardening and tempering heat treatment practice.

Further details on cast irons can be obtained by reference to the book by Elliot [5].

The number of materials available under the heading of steels is very large and they are generally classified under three headings:

plain carbon steels
low alloy steels (total alloy content < 8%)
high alloy steels (total alloy content >8%)

the latter group contains the heat, wear and corrosion resistant alloys.

***Plain Carbon Steels.*** Steels may be melted by induction or electric arc furnaces. If induction melting is chosen, there is a need to maintain close control over the composition of the charge materials as there is no opportunity for refinement. Electric arc melting, on the other hand, is ideal for refining and the term steelmaking is often applied when it is used. The approach adopted is based on the controlled oxidation of such elements as carbon, silicon and manganese. Steels are susceptible to defects and/or the impairment of properties caused by the failure to control the levels of oxygen, nitrogen and hydrogen. The latter two may be reduced to acceptable levels by the controlled oxidation of carbon. However, the residual oxygen must be removed by deoxidation if carbon monoxide gas hole formation in the castings is to be prevented. Silicon and manganese are mild deoxidisers, but aluminium is more effective and is commonly used as a ladle addition.

Heat treatment by hardening and tempering is possible for steels containing >0.4% carbon.

***Low Alloy Steels.*** The principal objective of alloying is to improve the hardenability of the steel, as this governs the depth to which hardening occurs in a section during quenching. Melting practice for this group of alloys is similar to that for the plain carbon steels.

***High Alloy Steels.*** Whilst the basic approach to melt processing is essentially similar to the previous groups, some of these alloys can benefit from special treatments. Melt cleanliness can be improved by an Argon Oxygen Decarburisation (AOD) treatment. In this refining treatment oxygen, argon and nitrogen are injected into the molten steel and produce a melt with reduced inclusions and gas content. Steels refined in this way exhibit significantly improved ductility and toughness, and use of the process has been extended to plain carbon and low alloy steels.

Niobium and titanium can be added to certain alloys and have the effect of stabilising the carbides. However, they must be added after deoxidation, otherwise their effect will be lost as the elements readily oxidise. Further information on steelmaking for steel founders can be found by reference to the book by Jackson and Hubbard [6].

## Superalloys

The superalloys are defined as those alloys developed for elevated temperature service, usually based on group VIIIA elements, where relatively severe mechanical stressing is encountered and where high surface stability is frequently required [7]. This definition includes iron-, cobalt- and nickel-based alloys. As the former group are covered by high alloy steels, only the cobalt- and nickel-base alloys are considered in this section.

***Cobalt-Base Alloys.*** These alloys demonstrate relatively low melting points, good fluidity, freedom from dissolved gas defects, and low alloy losses due to oxidation. The majority of the alloys are melted in air since they are devoid of such highly reactive elements as aluminium, tantalum, titanium or zirconium. However, inert gas (argon) covers are recommended to minimise oxidation, and vacuum melting is essential if the aforementioned reactive elements are present. Silicon and manganese can be added to certain alloys to improve alloy fluidity and act as deoxidants. Heat treatment has little effect on mechanical properties.

***Nickel-Base Alloys.*** These alloys are far more complex in their formulation than are the cobalt-base alloys. They are used in the most demanding applications and contain reactive elements to improve their mechanical properties and surface characteristics. Vacuum melting and casting is essential. Whilst no other melt treatment is required, a grain refining effect can be induced, in castings requiring an equiaxed structure, by incorporating a refining agent in the investment casting mould slurry formulation. The nickel-base alloys are amenable to heat treatment by solution treatment and precipitation hardening.

The special techniques of directional solidification and single crystal production have been developed for nickel-base alloy turbine blade production. This technology is featured in the latest military and civil aircraft engine designs. The technique of hot isostatic pressing was mentioned briefly in relation to titanium castings and is used for enhancing the performance of superalloy castings. These techniques are described in the following section.

## SPECIAL TECHNIQUES

### Directional Solidification

Turbine blades with an equiaxed grain structure tend to fail by creep fracture and thermal fatigue, with failure associated with those grain boundaries that are transverse to the applied stress. It was reasoned, therefore, that by eliminating such boundaries by producing a columnar grain structure parallel to the stress axis, grain boundary fracture could be eliminated and casting performance improved. This is the principle of the directional solidification process, which is an extension of the principle of directional solidification described in the previous chapter.

***Casting Production.*** A schematic diagram of the equipment for producing directionally solidified castings is shown in Fig. 10.6. A ceramic shell mould is attached to a water cooled chill. The assembly is surrounded by a furnace unit and the whole device is contained within a chamber which can be evacuated. When the assembly is at the required temperature molten alloy is poured into the mould. Solidification commences at the chill surface and is encouraged to progress directionally by slowly withdrawing the mould from the furnace to ensure that the desired grain structure is produced. Two conditions must be simultaneously satisfied. Firstly, the heat must be withdrawn from the casting in a unidirectional manner, and secondly, a positive temperature gradient must be maintained ahead of the solidifying interface in order to prevent extraneous nucleation.

Apart from the grain structure, the process also promotes casting soundness.

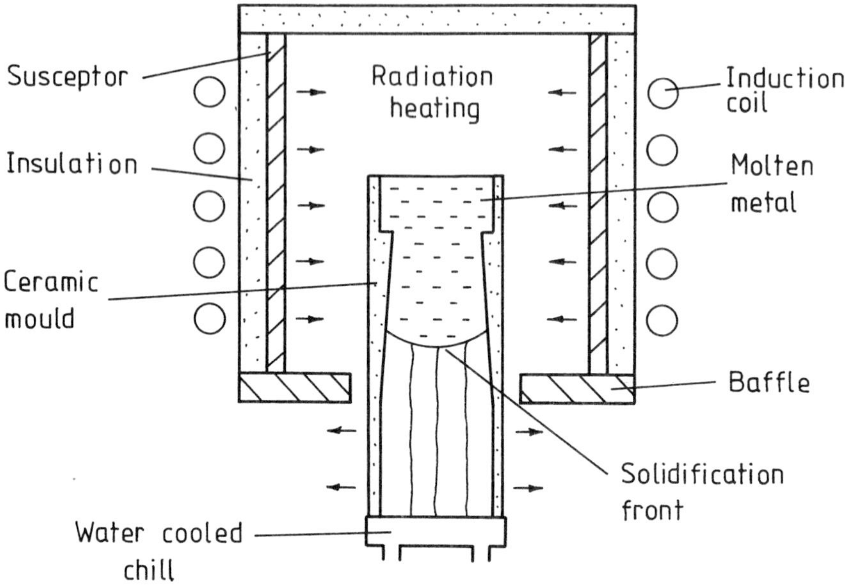


Fig. 10.6 A schematic representation of the equipment for producing directionally solidified castings

## Single Crystal Production

Single crystals have no grain boundaries and therefore do not require the presence of those elements added specifically to strengthen grain boundaries. As these elements lower the melting point of the alloy, their removal increases the alloy's melting point, and consequently increases the temperature at which it can be used in service. The equipment and techniques used for directional solidification can be applied to single crystal production. However, a modification in technique is required to ensure that only one crystal grows in the blade. This is achieved either by the use of a seed crystal or, more commonly, by the use of a special constrictor adjacent to the chill plate. The objective of the special constrictor is to permit only one grain to enter the airfoil portion of the mould. Whilst various designs of constrictor have been proposed, the helix appears to be preferred, see Fig. 10.7. Only one crystal survives growth through the constrictor to emerge into the blade form.

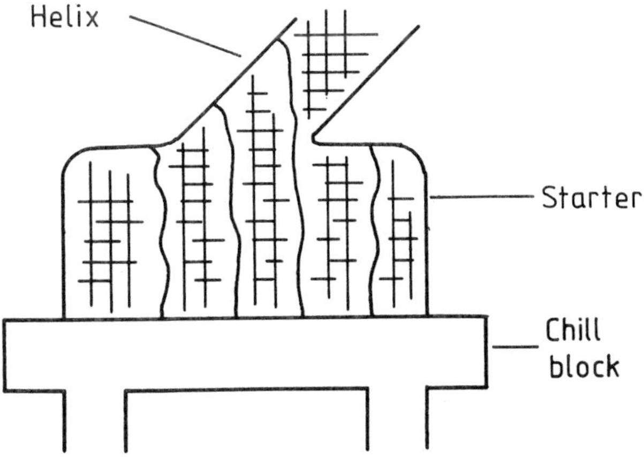


Fig. 10.7 The use of a helical constrictor to encourage single crystal growth

***Mould Material Requirements.*** In conventional investment casting the mould material is in contact with liquid metal for a very short period of time. However, in the directional solidification process there is a prolonged period of contact between the mould material and the metal. Clearly this procedure places much greater demands on the chemical and mechanical integrity of moulds and cores than does conventional casting. It may be necessary to introduce alumina into the shell building procedure to improve the mould's high temperature properties. The combination of alumina and a silica binder may result in the formation of mullite when the shell is fired and cast, thus improving refractoriness.

Further details of directional solidification can be obtained by reference to the book by McLean [8].

## Hot Isostatic Pressing (HIPing)

Hot isostatic pressing, or HIPing, is a process used to produce near net shape parts from metal, ceramic and cermet powders; to diffusion bond similar and dissimilar materials, including metals to ceramics; to improve the properties of parts formed by conventional powder compaction processes; and to heal microporosity in castings, thus improving their mechanical properties and extending their service life [9].

In the process, parts are subjected to high pressures and temperatures in a pressure vessel, see Fig. 10.8. An inert gas, usually argon, is the pressurising medium. For the processing of castings, argon is applied at pressures between 103 and 206 MPa (15,000 and 30,000 lbf/in$^2$). Process temperatures vary from 480°C (896°F) for aluminium alloy castings to 1315°C (2400°F) for single crystal nickel-base superalloys.

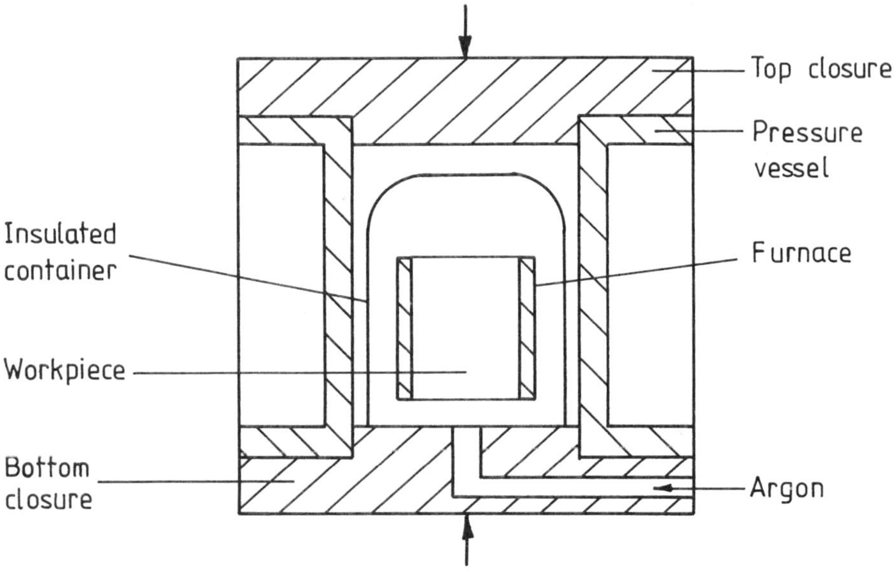


Fig. 10.8 Schematic diagram of the hot isostatic pressing process

When castings are processed, the simultaneous application of heat and pressure virtually eliminates voids and microshrinkage through a combination of plastic deformation, creep and diffusion. Since pressures are applied isostatically (uniformly in all directions), parts with intricate shapes, including those with thick and thin sections, can be processed without distortion and with uniform properties throughout. Dimensions are not normally affected since the voids are usually small.

The process cannot seal surface connecting defects and high purity argon is essential. Contaminants such as nitrogen, oxygen, methane, carbon dioxide, and hydrogen, even in trace amounts, may cause undesirable surface reactions. Modern equipment permits cycle times of between two and six hours, with cooling rates sufficiently rapid to allow the solution heat treatment of many alloys at the end of a cycle. With the increase in productivity, costs have reduced and the process is being used more routinely.

## References


1. Gilchrist, J. D. (1977). *Fuels, Furnaces and Refractories.* Published by Pergamon Press, Oxford.

2. Barber, H. (1983). *Electroheat.* Published by Granada Publishing Ltd, London.

3. Flemings, M. C. (1974). *Solidification Processing.* Published by McGraw-Hill Brook Company, New York.

4. Polmear, I. J. (1981). *Light Alloys: Metallurgy of the Light Metals.* Published by Edward Arnold Publishers Ltd, London.

5. Elliott, R. (1988). *Cast Iron Technology.* Published by Butterworth and Co (Publishers) Ltd, London.

6. Jackson, W. J. and Hubbard, M. W. (1979). *Steelmaking for Steelfounders.* Published by Steel Castings Research and Trade Association, Sheffield.

7. Simms, C. T., Stoloff, N. S. and Nagel, W. C. (1987). *Superalloys II.* Published by John Wiley and Sons, New York.

8. McLean, M. (1983). *Directionally Solidified Materials for High Temperature Service.* Published by The Metals Society, London.

9. Hanes, H. D. and McFadden, J. M. (1983). HIPing of castings: an update. *Metal Progress*, **123**, 5, 23-31.

CHAPTER 11

# Quality Control and Assurance

## INTRODUCTION

To receive an order for castings, a foundry must satisfy the customer that it can produce those castings at an acceptable cost, deliver them to an agreed schedule and ensure an acceptable standard of quality. The requirements for quality are established by the customer and are specified within national or international standards, which define the chemical composition and mechanical property ranges expected of the alloy selected; and by the component drawing, which defines the shape, dimensional tolerances and surface finish required. Furthermore, the customer has the right to expect that the castings will be sound, in that they will be free from internal defects such as gas or shrinkage porosity, and surface defects such as tears or cracks.

To comply with these requirements a foundry must implement effective quality control and assurance procedures. Such procedures commence with effective methods for checking the quality of incoming materials used for the production of castings. They are extended to the manufacture of castings, which includes both the mould production and metal melting stages, with the specific objective of eliminating process-related casting defects. Finally, the quality of the castings produced must be assessed to ensure that they meet the drawing standard and the requirement to be free of casting defects.

To meet these requirements a foundry has at its disposal a whole range of tests, from those used to determine the chemical, physical and mechanical properties of raw materials, mould materials and alloys, to those used for the non-destructive detection of internal defects in castings. It is the purpose of this chapter to identify the tests which are available and to briefly describe their procedures, capabilities and limitations.

The concept of quality assurance requires that quality should be built into the casting during the manufacturing process [1]. Indeed, quality should be built into the manufacturing process itself in the form of the design of the process and in the training of the personnel. However, quality castings begin with the designer, so that interaction between the customer and the foundry is fundamental to the production of quality castings [2]. Close control of all aspects of the manufacturing process is essential for quality casting production and the technique of statistical process control is invaluable in this respect. Details of this technique are also included within this chapter.

Quality assurance is cost-effective. It costs money to implement and maintain, but these costs are offset by the reduction in scrap and rework and the improvement in customer confidence.

## MOULD MATERIALS

For those mould materials based on the use of sand, a variety of established tests are used to monitor and control their quality. These tests include grain size distribution and grain fineness number, which determine the base sand's permeability, binder demand and influence on surface finish and are therefore important in the inorganic and organic bonded systems and even the unbonded sand systems. Beyond these tests the requirements will be different, depending upon the specific bonding agent. For the clay-bonded, green sand system, tests to establish the moisture content, active clay level, compactability, compression strength, deformation characteristics, and permeability of the sand are widely practised [3]. They may be supplemented by tests to establish volatile matter content or loss on ignition, which are important in relation to the level of organic material in the sand. These tests, developed essentially by the American Foundrymen's Society [4], provide the basis of sand testing and control, but may provide insufficient information for modern green sand moulding methods and to meet these requirements new tests have been developed [5-7].

The tests conducted on cold-set sands include those for strength in compression and tension. Such tests may be used to monitor the bench life and strip time during the production of moulds and cores, as well as the final strength on bond development or the effect of storage on strength changes. Other tests which are important for the effective control of these processes include those for temperature, in particular of the sand, and those for ensuring the quality in systems using reclaimed sands [3, 4].

In the case of the ceramic moulding processes, tests to establish the silica content, pH value and gelling time of the silicate binders are practised. These tests can be extended to the dipping slurries used for investment casting and can be supplemented by temperature, viscosity and specific gravity tests [8]. In the case of the fired moulds, control can be exercised by monitoring permeability and strength.

In the permanent mould processes the die material remains a constant, so that it is the die temperature, metal temperature and die filling procedures which are the factors to be controlled.

## METAL MELTING AND COMPOSITION CONTROL

The importance of the metal melting stage was described in Chapter 10. Regardless of the alloy being cast, metal temperature is an important variable to monitor and control. Incorrect control can result in a variety of problems such as the loss of volatile elements, absorption of gases, metal penetration of the mould when the temperature is too high, and misrun castings when the temperature is too low. The most popular method of measuring metal temperature is that based on the immersion of a thermocouple but optical and radiation pyrometers are also used.

Quality control tests to establish the level of gas in the molten metal are important for those alloys which are especially susceptible to gas absorption during melting. These include aluminium- and copper-base alloys, which are susceptible to hydrogen absorption. The tests available range from simple and effective qualitative methods, such as solidification of samples under reduced pressure followed by sectioning to identify gas porosity, to quantitative methods based on vacuum diffusion extraction of a gas from a solid sample followed by a measurement technique.

**Chemical Analysis**

The chemical composition of an alloy has a direct bearing upon the alloy's mechanical and physical properties, and is therefore closely defined within the alloy's specification. In most foundries there will be a facility for analysis, ranging from simple qualitative tests for the less sophisticated alloys to complex quantitative tests for advanced alloys used in safety-critical applications. The most meaningful results would be obtained by analysing the casting itself. However, this would be a destructive test and therefore acceptable alternatives must be used. For the most critical requirements, test coupons cast as part of the casting's gating system provide an excellent match to the casting's composition. For less demanding requirements, it may be sufficient to draw a sample from the ladle or furnace to provide a representative value for the composition. In this latter case the foundry must be sure that the metal is unlikely to lose volatile elements during the melt transfer and pouring stages.

***Qualitative Tests.*** In the case of cast iron, for example, it is possible to obtain a guide to the alloy's composition from two simple shop-floor tests. In the first, known as the chill test, a sample of metal is poured into a wedge-shaped cavity in a sand mould. The alloy solidifies quickly in the thinnest section as white iron. The lower the carbon and/or silicon content of the melt, the greater will be the depth of this chill layer. By calibrating the depth of chill against known compositions, the foundry has a rapid guide to a melt's suitability. In the carbon equivalent test a sample of cast iron is poured into a special cup containing a thermocouple. From the cooling curve it is possible to establish the iron's carbon equivalent value, which is a function of its carbon, silicon and phosphorus contents. This value correlates closely with the alloy's mechanical properties and provides a guide to the alloy's acceptability before pouring. The technique of thermal analysis can be extended to provide a good estimate of carbon and silicon content when the phosphorus content is consistent and low.

***Wet Analysis.*** The traditional method of analysis requires that the alloy be reduced to an aqueous solution using treatments appropriate for each element to be analysed. Gravimetric or volumetric techniques may then be used to establish an element's concentration to a high degree of accuracy. Such procedures are most effective when used to establish the major constituents present in an alloy [9]. The procedures require skilled chemists and are often time consuming.

***Spark Emission Spectrometry.*** In this technique a high energy spark vaporises metal from the surface of a prepared cast sample in an evacuated chamber. Diffraction gratings are used to separate the wavelengths of light emitted from the elements of interest and the amount of light energy emitted by each element is analysed and converted to a concentration value for that element. Properly standardised equipment can, with the incorporation of a powerful computing program, routinely provide the analysis of a sample within 5 minutes. The equipment is relatively expensive, but has the potential to provide an analysis for up to 48 elements.

***X-Ray Fluorescence Spectrometry (XRF).*** This technique utilises the interaction of the inner-shell electrons with primary X-rays [9]. The secondary X-rays produced are separated into the wavelengths related to the elements to be measured, using either energy dispersive or wavelength dispersive methods, and the intensity of each is recorded. Although expensive, the equipment is versatile and widely used in the non-ferrous industry.

## Mechanical Properties

In addition to composition, specifications for castings usually refer to expected values for mechanical properties. It is the normal practice to determine such values from separately cast test bars which are subsequently machined to specified dimensions before testing. It is important to clarify the status of the test bar to prevent any misunderstanding between supplier and customer. The test bar provides a guide to the properties of the castings. Its properties will be representative of the quality of the melt. However, because the test bar is produced to specific dimensions, it should not be expected to guarantee the properties in a casting of varied sections and subject to a different rate of cooling. Component specific data can, of course, be determined by removing sections from a casting for evaluation. However, this is a destructive test and would usually be confined to the design development stage.

The most commonly specified tests for mechanical properties are hardness and tensile tests [10].

***Hardness Tests.*** These tests usually measure the resistance to penetration by the casting surface and in a number of alloy systems the test provides a good correlation with strength and wear resistance. The Brinell hardness test, which is popular for non-ferrous alloys and cast irons, employs a hardened steel ball indentor. The Vickers test employs a diamond pyramid indentor and the Rockwell test a diamond cone. The latter test is the most flexible and the easiest to use in manual or automated procedures.

***Tensile Tests.*** In this test the specimen is subjected to a tensile load until it fails. It is a popular test because it provides information on the strength of the alloy, in the form of tensile strength and yield or proof stress, and ductility, in the form of elongation and reduction in area values.

Other tests which may be applied to cast material include: impact testing, fracture-toughness testing, and fatigue testing. Impact tests include the Charpy and Izod tests. The former is used to determine the ductile to brittle transition temperature of the alloy and provides excellent empirical correlation with service experience. The Izod test is conducted at room temperature. Fracture-toughness testing uses the concept of fracture mechanics to predict the load-carrying capabilities of components containing cracks. Fatigue testing recognises that components subject to cyclic loading at a level below the yield stress may fail after a certain number of cycles. Many cast components are subject to cyclic loading in service and it is believed that some 90% of all service failures due to mechanical causes arise from fatigue.

## CASTING EVALUATION

The evaluation of a casting's quality usually commences with a visual inspection which will reveal the presence of any immediately obvious defect [11, 12]. However, visual inspection cannot be guaranteed to detect fine surface cracks, nor is it appropriate for establishing that a casting is within dimensional tolerances and free from internal defects. The dimensions of a casting can be established by conventional metrology [12], or castings can be classified as acceptable or not by the use of gauges, jigs or templates. Whilst surface roughness values have been quoted for the various casting processes described, it would be fair to state that such measurements are not routinely conducted on the surface of castings. Objective values for surface roughness are most easily obtained for castings produced by investment casting or pressure die casting. However, for castings produced by sand moulding processes subjective evaluation, possibly assisted by reference to a set of standard surfaces, is more likely.

Castings which are required to contain fluids can be tested for leaks either by pressurising them with air and observing the emission of bubbles when they are immersed in water, or by observing the leakage of a pressurised penetrating liquid [10]. However, the most important group of tests used to evaluate a casting's quality are those described as non-destructive tests.

Non-destructive testing (NDT) provides information about the quality of a casting without causing any change in the nature or structure of the castings tested. These tests can be considered under three headings:

Tests to detect surface defects.
Tests to detect internal defects.
Tests to check structure.

## Tests to Detect Surface Defects

A visual inspection will not be effective in identifying the presence of small tears or cracks in the surface of a casting. The presence of such cracks will be especially deleterious to the fatigue life of the component. Methods are therefore required to positively identify the presence of such defects so that unacceptable castings are not delivered to the customer.

***The Magnetic Particle Flaw Detection Test.*** This test relies upon the ability to magnetise the casting under test and is therefore only suitable for ferromagnetic materials. The principle of the test is very simple, see Fig. 11.1.

A discontinuity in the surface of a magnetised casting, such as a crack, will create magnetic flux leakage. The presence of this leakage can be detected by applying finely divided, magnetisable particles to the surface of the casting. These particles are attracted to the point of flux leakage and thus indicate the location, size and shape of the discontinuity. The methods used to magnetise the casting are [11]:

- the flow of the electric current from point to point in a component by the use of prods (this is suitable for the inspection of large castings).
- the use of a coil carrying an electric current, the coil encircling all or part of the component (this is suitable for smaller castings and would be the basis of automated testing).
- the use of a permanent or electromagnet (this method is appropriate when there is no electricity supply available).

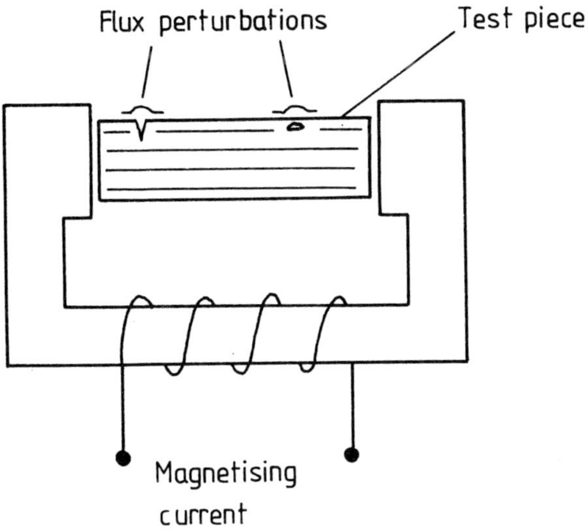


Fig. 11.1 Schematic diagram of the magnetic particle flaw detection test

The magnetic particle flaw detection test is the most reliable test available for detecting cracks at the surface of ferromagnetic iron and steel castings [12], with good sensitivity for detecting small flaws. This sensitivity is enhanced by the use of liquid fluorescent inks for examination under ultra-violet light because of the improved contrast provided.

Although the test provides the greatest sensitivity when the discontinuity breaks the surface at right angles to the flux, this restriction has been reduced by test equipment in which magnetic fields can be applied in two or more directions. Whilst the procedure can be automated, interpretation is still operator dependent. However, attempts are being made to use image processing techniques to overcome this limitation [13].

***The Penetrant Flaw Detection Test.*** This test is suitable for ferrous and non-ferrous alloy castings and will detect defects providing that they communicate with the surface. The test relies on the capillary action of a crack which will draw liquid into it. The subsequent leakage of this liquid may be detected by suitable means and the location of the defect identified. Two types of penetrant are used [11]:

- dye penetrants, which are liquids containing a coloured dye.
- fluorescent penetrants, which are liquids containing a substance that will emit a green/yellow light when illuminated with ultra-violet light.

The penetrants may be carried in either water or an organic liquid and are sprayed over the surface of the component, with the excess removed by washing in an appropriate liquid. When a dye penetrant is used a developer is applied to the surface, typically a suspension of chalk in a volatile liquid carrier. Sufficient time is then allowed for the penetrant to seep out of the cracks and stain the porous developer, thus revealing their positions. When a fluorescent penetrant is used the development stage can be omitted and the presence of cracks detected using ultra-violet light. Highly automated systems, including defect recognition vision systems, are available for the fluorescent dye penetrant test [12].

***The Eddy Current Test.*** When a coil carrying an alternating current is placed in proximity to a conductive material, secondary or eddy currents will be induced within the material [14]. The induced currents will produce a magnetic field which will be in opposition to the primary magnetic field surrounding the coil. This interaction between fields causes a back emf in the coil, and hence a change in the coil impedance value. If a casting contains a crack or other discontinuity this will change the flow patterns of the eddy currents, the magnetic field, and hence the coil impedance. As the impedance of a coil can be determined by measuring the voltage across it, a simple detection device can be used to display the presence of a defect. Whilst this method may detect surface cracks under specific conditions, limitations arise because of surface roughness and variations in structure [11].

## Tests to Detect Internal Defects

Radiographic and ultrasonic methods are well established for the detection of internal defects in castings and are likely to remain dominant for the foreseeable future [12].

***Radiography.*** Radiography is based upon the principle that penetrating radiation is differentially absorbed by the material through which it passes. The absorption is primarily dependent upon the energy spectrum of the radiation and the density and thickness of the material [15]. The radiation which passes through the material can be detected by one of several methods to provide information about the material. The radiation emerging from a casting of uniform section thickness should itself be uniform. If there was a defect in that section, such as a shrinkage cavity, then the reduced thickness of the casting at that point would permit more radiation to emerge and this difference could be detected. Volumetric defects, such as gas or shrinkage cavities and inclusions, are readily detected by radiography whatever the plane of the incidence of the radiation. However, planar defects, such as cracks, are more difficult to detect and their detection depends upon the direction of the radiation.

Radiation in the form of X-rays is generated when a beam of high energy electrons impinges on the surface of a metal target (anode). The sudden deceleration of the electrons generates X-rays. A schematic diagram of an X-ray generator is shown in Fig. 11.2. These generators are limited to a tube voltage of 1 MV, which would provide radiation capable of penetrating up to 130 mm (5 in) of steel. High energy sources of electron beams include the betatron, Van der Graaf generator or linear accelerator. The latter is the more common and equipment with the power to penetrate 400 mm (16 in) of steel is available. An alternative source of radiation is found in γ-rays, which are generated during natural radioactive decay. The rate of decay and the energy of the γ-ray are important parameters. The two most commonly used sources of γ-radiation are Iridium 192 and Cobalt 60 [15].

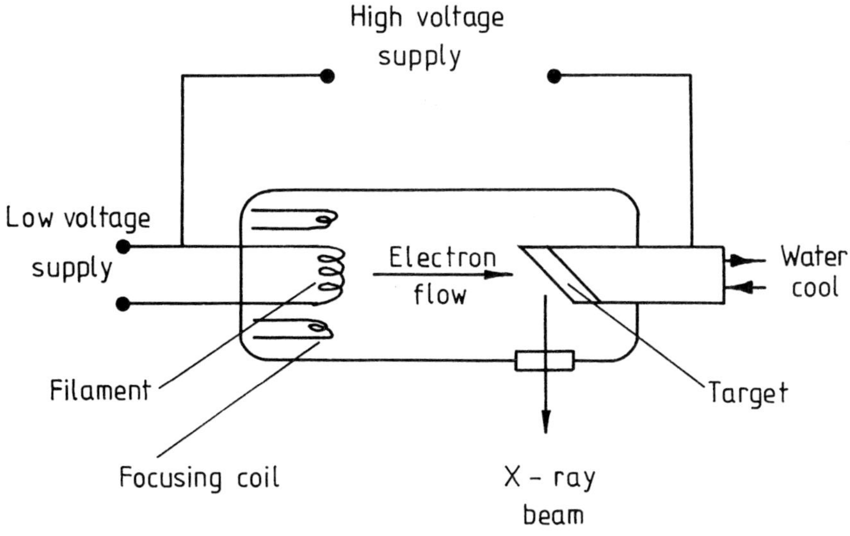


Fig. 11.2 Schematic diagram of an X-ray generator

Whilst the use of photographic film, processed manually or automatically, to provide a permanent image for evaluation and storage still predominates, alternative techniques of increasing sophistication have been developed [16-18]. Real-time radiography using fluoroscopy is well established and provides such benefits as reduced costs, improved productivity and improved defect detection [19, 20]. It is possible to record the images, using video techniques, to provide a permanent record. The interpretation of radiographic images has always required skill and represents a potential source of error. With the use of digital electronics and computers it is possible to enhance real-time images, or conventionally produced radiographic film images, to enable features not normally discerned by the human eye to be detected [15, 17].

Although currently very expensive, computed tomography (CT), also known as computerised axial scanning (or CAT scanning), is being used for cast turbine blade inspection [12, 16]. This X-ray imaging technique was originally developed for medical diagnostic use. It involves moving the detector or component in a stepwise fashion through 360°. The images generated are processed by computer to build up a composite picture of the section being examined.

***Ultrasonics.*** Sound waves are mechanical waves which can be transmitted through fluids and solids. The term ultrasonic is used to describe sound with a frequency beyond the range audible to a human being, which is typically above 20 kHz. The frequency range employed in the inspection of castings is 0.5 to 5.0 MHz, the predominant factor being the relationship between the wavelength of the sound and the grain size of the cast structure. Not all cast alloys can be ultrasonically examined [11]. Probes are used to transmit and receive the ultrasound. A reflection is received from an interface, such as the back wall of the casting or the surface of a defect. This information can be displayed on a cathode ray tube with the time base calibrated to represent the thickness. Material of a given thickness will display a trace in which the position of the back surface echo is a function of the time taken for the sound wave to pass from the transmitter, through the material to the back wall, and back again to the receiver. If there is a defect present in the section then the echo from that defect will be received earlier than the echo from the back wall. This will show as an additional feature on the oscilloscope trace and its position, in relation to the time base, can be used to determine the depth at which the defect occurs within the section. This method of information presentation is referred to as the 'A' scan and is shown schematically in Fig. 11.3. A disadvantage of the 'A' scan is that there may be no permanent record, unless a photograph is taken of the screen image, although the more sophisticated modern equipment has the facility for digital recording [14]. With the development of 'B' scan and 'C' scan displays associated with semi-automatic or automatic testing techniques and digital information processing, permanent records of the position of defects within a material can be obtained.

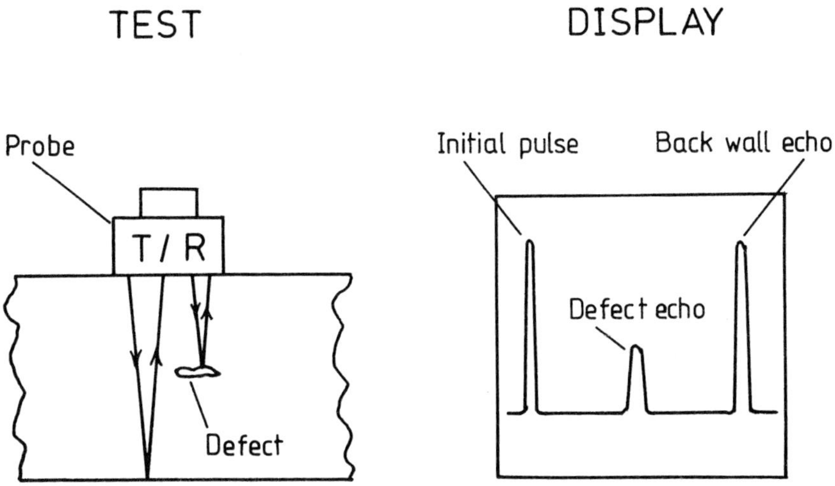


Fig. 11.3 Schematic representation of the ultrasonic method of non-destructive testing

Conventional piezo-electric transducers are mechanically coupled to the surface of the component under test. However, alternative non-contacting methods are available for exciting an ultrasonic wave and include the use of electromagnetic acoustic transducers and pulsed laser generation [21].

Other NDT techniques which have the potential to locate defects in castings include acoustic emission and laser holography, both of which require the component to be examined under stress [12]. With acoustic emission the high frequency noise generated by the deformation of material under stress is monitored using probes attached to the surface. Emissions can result from the propagation of cracks, plastic deformation and from phase changes. Holography uses laser technology to produce interference fringe patterns on the surface of a stressed component. The stress around an internal defect causes a change in the surface fringe patterns, thus enabling the presence of a defect to be detected.

**Tests to Determine Structure**

A number of techniques are available for an assessment of the structure of castings. The eddy current test, briefly described previously with respect to defect detection, can be used to assess structure. The test relies on the comparison of a test casting with a standard casting. If the structure of the test casting deviates from that of the standard casting, then there will be a variation in the level of eddy current induced. Detection of this variation can be used as a method for separating castings with unacceptable structure. This test procedure has been used successfully in the differentiation and sorting of as-cast and heat treated castings [12, 13, 22].

Sonic testing has been used to determine the type and amount of graphite present in cast iron components [11]. The physical properties of cast irons, such as elastic modulus and density, are affected by the type and amount of graphite present and by the condition of the matrix. The test involves the measurement by electrical means of the resonant frequency of a casting. This resonant frequency will be proportional to the modulus and density, and therefore proportional to graphite form. Like the eddy current test, this test relies on a comparison being made between an acceptable standard casting and the casting under test.

Ultrasonic velocity measurements are widely used as a means of guaranteeing the nodularity of the graphite structure in SG iron castings and, if the matrix structure is known to be consistent, to guarantee the principal material properties [12, 23]. Measurement of the attenuation of ultrasonic energy over a wide frequency band has been used to evaluate structure and grain size in materials. With further development to improve sensitivity and to enable reliable tests to be carried out on castings, attenuation measurements could prove of considerable value.

A real-time X-ray diffraction technique has been developed by Rolls-Royce plc for checking the crystal orientation in single crystal turbine blades [18]. Known as SCORPIO, the technique is highly automated and uses a robot to position the blade in the detector system. The orientation of the structure in the test blade is compared to a computer-generated standard and the test blade is rejected if its structure is outside the acceptable tolerance limits.

## STATISTICAL PROCESS AND QUALITY CONTROL

In the production of castings by any casting process there are a large number of variables which can exert an influence on the quality of the casting. The output of any process, casting or otherwise, will be subject to variations. However, providing that these variations lie within an agreed range determined by the customer or contained within a recognised standard, the product will be acceptable. It is therefore essential that any process be controlled so that any variations in that process do not result in unacceptable variations in the final product. The objectives of statistical process control (SPC) are: firstly, to ensure that a process is in control by removing all special (non-random) causes of variation, so that the results of the process can be assured; and secondly, to identify those contributors to random process variation which can be tackled to reduce historically accepted random process variation. In this respect SPC provides a tool for the continuous improvement of a process and, therefore, the quality of the product.

Measuring the quality of a casting, using whichever variables are important, will indicate whether or not the process is in control. However, the feedback provided by this approach delays any action to change the process to return it into control, should it be deviating. To obtain an improved response time, the variables of the process itself must be monitored so that immediate feedback is provided to control the process. In practice, information from both sources, the product and the process, will be used. The important factor is that this information be analysed quickly so that corrective action can be taken to minimise the production of an unacceptable product.

A detailed consideration of SPC is beyond the scope of this book and the reader should consult specialist publications [24, 25]. However, in order to explain the methodology of SPC, the procedures for establishing and using control charts will be described [26].

The basis of SPC is the normal distribution curve shown in Fig. 11.4. This curve records the number of times (frequency) that a particular value of a variable occurs. It will be noticed that the majority of values are clustered around the centre of the curve and that the mean (average) value ($\bar{x}$) corresponds to the peak of the curve. When a distribution appears normal it is possible to calculate the standard deviation ($\sigma$), which is a universal index of variation. It expresses standard amounts of variation regardless of the units of measurement. For a normal distribution, 68.3% of the measured values will fall within $\pm\ \sigma$ of the mean, 95.5% within $\pm\ 2\ \sigma$, and 99.7% within $\pm\ 3\ \sigma$. This provides the basis for the construction of a statistical process control chart.

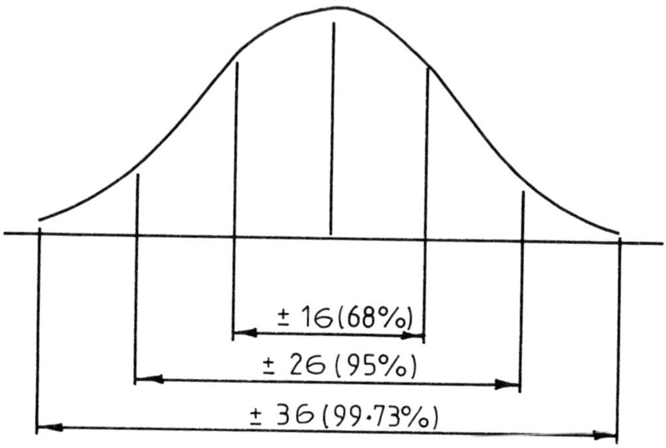


Fig. 11.4 The normal distribution curve

## SPC Charts

To enable a chart to be constructed, sufficient data for the variable concerned must be collected under normal production conditions, where only random causes of variation are present, for which it is known that the limits of the process variation are acceptable. Data is usually collected in the form of samples of consecutively produced components. From the data, values for the mean and standard deviation can be estimated by statistical calculations. This data is used as the basis for obtaining control limits on control charts. The two most common types of control chart that are used are the average ($\bar{x}$) and range (R or W) charts. An example of the former is shown in Fig. 11.5.

Data collected from the process in the form of averages and ranges of small samples of consecutively produced product may now be plotted on the chart against time. Under normal conditions the data will be distributed randomly within the envelope delineated by the lower and upper warning limits.

However, there is approximately a 1 in 40 chance that a value will fall outside the warning limits and approximately a 1 in 1000 chance that a value will fall outside the action limits. Corrective action will, of course, be essential in this latter situation. However, the chart provides a visual indication of trends which may indicate that the process is moving out of control. This enables early corrective action to be taken. Examples of warning signs to look for include [27]:

- one or more points fall outside either the lower or upper control limit.
- there are more than six consecutive points on one side of the mean.
- the trend in the data is either up or down.
- more than half the points are above or below the mean.
- there is an obvious non-random pattern present.

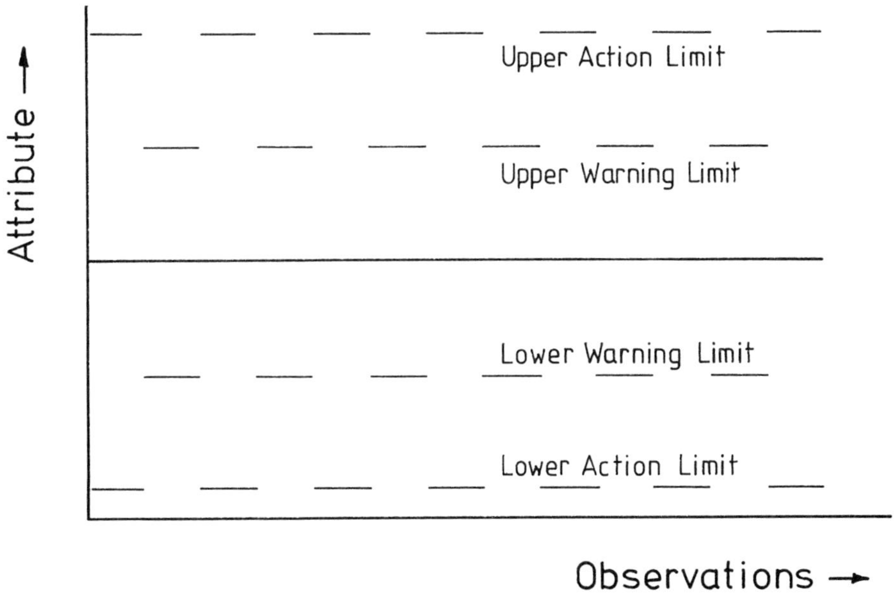


Fig. 11.5 A process control chart

Providing that the data falls within the lower and upper warning limits, the process can be considered to be in control and the product acceptable. This effectively defines the process capability. Under these conditions the first objective of SPC has been attained. However, the second objective of SPC is to seek continuous improvement in the process, which means a reduction in the variation of the product produced by the process. How can this be achieved?

To illustrate the answer to this question we can consider, as a specific example, the production of green sand moulded castings.

The dimensional accuracy of a green sand moulded casting is a function of a number of variables, such as the pattern and moulding equipment accuracy, the mould material properties, and the metal composition. Assuming that the method used for producing the moulds is an air actuated, jolt-squeeze moulding machine, then the maximum density or hardness of the mould will be a function of the air line pressure. Clearly there are a number of variables which interact to determine the process capability. However, to demonstrate the point, we shall consider variations in moisture content of the sand. At higher moisture contents the sand will be more deformable and, because maximum mould hardness cannot be obtained by the moulding method, the metal can deform the mould to produce a slightly oversize casting. However, under normal conditions, when the process is under control, this variation in dimensions will be accepted as within the process capabilities. The customer accepts this variation as normal.

If the customer introduces new equipment for machining castings that requires the castings to be more consistent than the process is normally capable of producing, then the foundry has a problem. It is possible, by improving the process, that the normal variation can be reduced to an acceptable level. For example, the existing process may operate with a moisture content of 3% ± 0.5%. If, by improved control, the range can be reduced to 2.75% ± 0.25% then, all other factors being equal, the variation in casting dimensions may be reduced to a level acceptable to the customer. However, should improvements in the existing process fail to provide castings acceptable to the customer, then it may be necessary to change the process, for example, to introduce a high pressure green sand moulding process.

SPC can be regarded as an effective, cheap and practical means of preventing a deterioration in overall quality. However, control charting itself does not reduce costs or improve quality, there must be effective feedback and corrective action [26].

## References


1. Ohyama, S. (1988). Total quality assurance - its concept and implementation in the foundry. *Proceedings of the BCIRA Conference entitled: Ensuring Quality Castings - Demands and Achievements Worldwide, UK,* Paper No. 7, 22 pp.

2. Long, L. H. (1981). Achieving quality casting production - part 1. *Foundry*, **109**, 12, 42-49.

3. Bailey, R. (1983). Quality through testing: part 1 - sand. *Modern Casting,* **73**, 6, 19-22.

4. *Mold and Core Test Handbook.* Published by the American Foundrymen's Society, Des Plaines, Illinois, USA, 1989.

5. Boenisch, D. and Ruhland, N. (1988). New concepts of green sand technology - part 1. *Foundry,* **116**, 2, 21-27.

6. Boenisch, D. and Ruhland, N. (1988). New concepts of green sand technology - part 2. *Ibid,* **116**, 3, 38-49.

7. Graham, A. L. (1985). Sand testing - a guide to system design. *Ibid,* **113**, 1, 48-55.

8. *Wax, Binder and Slurry Test Handbook.* Published by the American Foundrymen's Society, Des Plaines, Illinois, USA, 1988.

9. Guy, S. (1987). Chemical analysis - its role in SPC. *Metals and Materials,* **3**, 4, 210-212.

10. Kanicki, D. P., Kotzin, E. L. and Lobenhofer, R. W. (1983). Quality through testing: part 3 - metal. *Modern Casting,* **73**, 8, 23-25.

11. Lavender, J. D. (1983). Non destructive testing and control. *Foundry Trade Journal,* **154**, 3261, 489-504.

12. Rickards, P. J. (1988). Progress in guaranteeing quality through non-destructive methods of evaluation. *The Foundryman,* **81**, 5, 196-209.

13. Reader, K. F. (1990). Automated NDT for manufacturing industry. *Metals and Materials,* **6**, 10, 636-639.

14. Hull, B. and John, V. (1988). *Non Destructive Testing.* Published by Macmillan Education Ltd, London.

15. Hamilton, J. W. and Hayton, J. A. (1985). An introduction to industrial radiography. *Metals and Materials,* **1**, 2, 108-112.

16. Bralower, P. M. (1986). Nondestructive testing - part 1: the new generation in radiography. *Modern Casting,* **76**, 7, 21-23.

17. Krohn, B. R. and Isaacson, B. G. (1988). Digital radiography: an NDT solution for casting defects. *Ibid*, **78**, 2, 24-26.

18. Heine, H. J. (1989). Automated NDT inspection of intricate precision castings. *Foundry*, **117**, 1, 37-40.

19. Lee-Frampton, J. (1982). X-ray fluoroscopic inspection of castings. *Foundry Trade Journal,* **153**, 3244, 227-248.

20. Brant, F. (1990). Casting defect detection using real-time radioscopy. *Foundry*, **118**, 5, 56-60.

21. Hudson, B. (1984). Horizons in non-destructive examination. *Foundry Trade Journal,* Quality Testing Supplement, December, S3-S6.

22. Beard, D. R. (1980). Assessing casting quality. *Engineering*, **220**, 11, 1247-1250.

23. Robinson, N. A. (1982). Recent advances in the non-destructive evaluation of cast irons. *Foundry Trade Journal,* **154**, 3244, 242-246.

24. Caplen, R. (1988). *A Practical Approach to Quality Control.* Published by Business Books Ltd, London.

25. Stebbing, L. (1989). *Quality Assurance: The Route to Efficiency and Competitiveness.* Published by Ellis Horwood Ltd, Chichester.

26. Wootton, R. and Knight, D. F. (1989). Quality assurance through the dynamic control of casting quality. *The Foundryman,* **82**, 11, 527-537.

27. AFS Green Sand Molding Committee (4M) (1989). SPC for small to medium foundry's sand system: part 3. *Modern Casting,* **79**, 11, 42-44.

# Index